SCORCHER

Published by **Black Inc. Agenda**
Series Editor: Robert Manne

Other books in the Black Inc. Agenda series:

Whitewash: On Keith Windschuttle's Fabrication of Aboriginal History ed. ROBERT MANNE

The Howard Years ed. ROBERT MANNE

Axis of Deceit ANDREW WILKIE

Following Them Home: The Fate of the Returned Asylum Seekers DAVID CORLETT

Civil Passions: Selected Writings MARTIN KRYGIER

Do Not Disturb: Is the Media Failing Australia? ed. ROBERT MANNE

Sense & Nonsense in Australian History JOHN HIRST

The Weapons Detective ROD BARTON

SCORCHER

THE DIRTY POLITICS OF CLIMATE CHANGE

CLIVE HAMILTON

WITH RESEARCH ASSISTANCE FROM
CHRISTIAN DOWNIE

Black Inc. Agenda

CONTENTS

Acknowledgments

1. THE GREENHOUSE MAFIA
2. FRAMING THE DEBATE
3. AUSTRALIA'S EMISSIONS
4. BOLD DECLARATIONS
5. THE ROAD TO KYOTO
6. DRAMA AT KYOTO
7. VICTORIES AND DEFEATS
8. AUSTRALIA: AFTER KYOTO
9. BUSINESS REALIGNMENTS
10. THE BURNOUTS
11. THE BATTLE FOR PUBLIC OPINION
12. COMICAL IAN
13. NEW GLOBAL MOMENTUM
14. THE TURNING TIDE
15. SABOTAGING THE FUTURE?

Notes
List of Abbreviations
Index

ACKNOWLEDGMENTS

Many people have contributed to this book in one way or another. Christian Downie has provided superb research assistance, which has made the task much easier and improved the final product considerably.

I am especially grateful to five people, each with expertise in aspects of the climate change debate, who read and commented on the entire manuscript: Peter Christoff, Andrew Macintosh, Barry Naughten, Hugh Saddler and Tony Weir. It is a much better book as a result of their comments.

Over the years I have written a number of papers with other authors. Some of the thoughts and materials developed in those collaborations have found their way into this book. In this context I would like to express my gratitude to Richard Denniss, Mark Diesendorf, Paul Pollard, John Quiggin, Justin Sherard, Alan Tate, Lins Vellen and George Wilkenfeld. Hal Turton in particular has written a number of papers with me or for the Australia Institute that have been of great benefit, as the text will show.

Perhaps my greatest intellectual debt is to Hugh Saddler, who has been a constant source of insight and good sense.

None of those mentioned above should be held responsible for anything that appears in the published version.

While I have referenced assiduously the various claims in this book, in some cases information has been communicated to me informally by officials, former officials or business-people who may suffer if their names are revealed.

Over the years a number of other people have contributed

substantially to my understanding of aspects of climate change in ways that have influenced this book. They include Bob Burton, Bill Hare, Catherine Fitzpatrick, Don Henry, Erwin Jackson, David Karoly, Graeme Pearman, Barrie Pittock, Tony McMichael and Cathy Zoi.

Finally I would like to thank Robert Manne, who suggested the project and provided advice, and Chris Feik at Black Inc. for his excellent editorial advice and support.

1. THE GREENHOUSE MAFIA

THE POLICY FIX

The inner workings of how climate change policy is actually decided in Canberra under the Howard Government were exposed by the ABC's *Four Corners* program in February 2006. We learned that for a decade the Government's policies have been determined by a cabal of powerful fossil-fuel lobbyists representing the very corporations whose commercial interests would be most affected by any move to reduce Australia's greenhouse gas emissions. In democratic societies, governments are supposed to represent the interests of the people. As will become clear, in the case of greenhouse policy the Howard Government represents the interests of a small but powerful group of corporations.

The story has been uncovered by the author of a doctoral dissertation completed in 2005 at the Australian National University. Guy Pearse, a member of the Liberal Party and a former adviser to Senator Robert Hill when he was environment minister, managed to coax the leading members of the fossil-fuel lobby into frank admissions about how they go about their business.[1] It emerges that climate change policy in Canberra has for years been determined by a small group of lobbyists who happily describe themselves as the 'greenhouse mafia'. This cabal consists of the executive directors of a handful of industry associations in the coal, oil, cement, aluminium, mining and electricity industries. Almost all of these industry environment movement on climate change. Emboldened by their success, 'they pursue the greenhouse agenda with an almost religious zeal', he wrote. According to the insider quoted above, the big early win that set the mafia on their victorious path through Government and the bureaucracy was when they combined in 1996 and 1997 to derail attempts by Phillip Toyne to promote a carbon tax.[6] Toyne was appointed by the Keating Government as deputy secretary of the environment department after a successful period as executive director of the Australian Conservation Foundation.

It was not just the environment groups that felt the power of the greenhouse mafia. Other industry groups that had a stake in measures to reduce greenhouse gas emissions – like the gas industry, the tourism industry and the insurance companies – were intimidated and went 'missing in action' for years. In bodies such as the Business Council of Australia, the hard-heads from the mining, coal and aluminium companies insisted that greenhouse was their issue and others should stay out. Pearse concluded that 'the intimidation or scarecrow effect of the greenhouse mafia has been central to the "missing-in-action" phenomenon'.[7] Nor was it only the greenhouse mafia (who at times also refer to themselves as 'the Society for Egomaniacs') who would give grief to any business that did not toe the line; so would the Government itself. When one senior businessman was asked why his corporation was unwilling to publicly urge the Government to ratify the Kyoto Protocol, he said that ministers made decisions affecting his company's commercial interests every week and he did not want to see the decisions start to favour his competitors.[8]

The Howard Government has allowed the greenhouse mafia extraordinary influence over Australia's stance on climate change. Alone among the nations of the developed world, Australia has included key members of fossil-fuel lobby groups in its official delegation to negotiate the Kyoto Protocol. Even the Bush Administration does not permit this unseemly arrangement, relegating fossil-fuel lobbyists to the gallery with the other NGOs rather than having them at the conference table. Said an insider: 'They are part of the [Government's] team. It is probably the best cross-industry alliance – the most successful ... of any one that has been put together ... We all write the same way, we all think the same way, we all worked for the same set of ministers'.[9]

Green groups have been no match for such a potent opponent when it comes to crucial policy decisions. This is when the inside knowledge and connections of the greenhouse mafia really make a difference, and when the democratic process is overruled. Cabinet deliberations, ministerial committees and preparation of Cabinet submissions are meant to be confidential and beyond the reach of lobbyists; indeed, the unauthorised disclosure of Cabinet-in-confidence materials is a crime. Yet Pearse's research reveals that the greenhouse mafia has unrivalled access to internal Government processes. Members of the greenhouse mafia even admit to being called in to Government departments to vet and help write Cabinet submissions and ministerial briefings, referring to 'mutual trust' between the lobbyists and the public servants (whose seats the lobbyists once warmed).[10] They have used this access to help public servants in the industry and resources department write submissions designed to counter proposals coming to Cabinet from the Australian Greenhouse Office through the environment minister. One of the most insightful industry insiders interviewed by Pearse made the following comments:

> I used to read the cabinet papers, you know, I know what was going on. And it was a question of using those 'ins' carefully and protecting sources, and you never go public on it. It is about fixing the outcomes. But at times there are some quite critical political games going on in there – power games,

information games ... [Environment minister] Robert Hill was sending letters to the Prime Minister trying to say that certain actions that he was about to take were consistent with previous cabinet decisions and so on, and the Prime Minister is about to go overseas and we get wind of it – we get a copy of the bloody letter, and then show it to Anderson [National Party leader and deputy prime minister] and other people on the ministerial committee who Robert hadn't bothered to copy in. And then, you know uproar breaks out because what Hill was trying to do was to slide it by the Prime Minister ... Robert Hill had a lot of trust with the Prime Minister after 1997 [the Kyoto conference], but then he started to over-play his hand and in the end he lost the trust of the ministerial committee and the backbench because he was seen to be arguing his own book.[11]

In this way the process of government is corrupted.

The greenhouse mafia has direct access to the Prime Minister. Said one: 'there's this arrangement where senior business people can ring Howard direct'. The resources company Woodside's huge investment in the North West Shelf had given its boss John Akehurst 'enormous influence over Howard'.[12] One celebrated incident at a meeting of the Minerals Council of Australia involved David Buckingham, a former senior bureaucrat in the environment department who became the executive director of the Business Council of Australia (BCA). Under the influence of Hugh Morgan and a handful of powerful mining and aluminium companies, the BCA took a strongly anti-Kyoto view that Buckingham wanted to soften. In arguing his case, Buckingham reportedly said that he had it from 'the highest levels of government' that industry ought to take a certain view. At that point Dick Wells, the executive officer of the Minerals Council, excused himself, left the room and, it's claimed, rang Arthur Sinodinos (Prime Minister Howard's then chief of staff). As one insider explained:

> So Dick calls Arthur – and he said, 'Arthur, Buckingham is sitting in a room next to me in my office here telling us that the government wants us to do this, this, and this. And he is talking like it is coming from you'. And Arthur says, 'Well, it has not come from me, and we do not want you to do it'. And so Dick walked back in and said, 'Look, sorry David – I just talked to Arthur Sinodinos and he disagrees completely with what you just said'. It was that sort of game.[13]

If early intervention failed and a proposal to tackle greenhouse gas emissions managed to get to Cabinet – such as occasionally happened when Robert Hill thought he could get something up – the mafia would turn to its closest friends in Cabinet to knock it off. Said one: 'if we wanted to put a spoke in the wheel of Robert Hill or whatever, we could do it pretty quickly … we reverse-managed that ministerial [greenhouse] committee so many times'.

When the inner workings of the greenhouse mafia and its influence on the Howard Government were exposed on *Four Corners*, the main players were indignant. Within hours, the coal lobby 'resolutely denied' obtaining undue access to Government decision-making processes. The executive director of the Australian Coal Association, Mark O'Neill (a prominent member of the mafia), attacked the program as a 'new low-water mark in shoddy, biased journalism' and denied they had been given any untoward privileges.[14] Pearse was dismissed as a 'self-styled whistle-blower with declared political ambitions'. The Murdoch press's most rabid commentator, Andrew Bolt, presented the Government's response, writing that the claims were mere hearsay and, besides, had been denied by the Coal Association.[15] In fact, he wrote, the 'real' greenhouse

University scientists around the country have been made aware that criticism of the Government jeopardises research funding.[19] Renewable energy researchers, some of whom still receive federal funds despite severe cuts, have been silenced. According to Philip Jennings, professor of energy studies at Murdoch University, renewables researchers believe they will lose their research funding if they are seen to criticise Federal Government policies on climate change and energy.[20]

Australian climate scientists feel under siege abroad as well as at home. Dr Tony Haymet succeeded Pearman as chief of the CSIRO Division of Marine and Atmospheric Science until September 2006, when he was appointed to head the prestigious Scripps Institution of Oceanography in San Diego. Noting the strong reputation of Australian scientists in climate change research, he said on his departure:

> Scientists are social animals, so they tend to generalise about scientists from particular countries, just as we did about scientists from the former Soviet Union … We are no longer perceived as being supporters of the UN or the world scientific enterprise. To meet someone for the first time and have them badmouth Australia, whether justified or not, is not easy.[21]

How has it come to this? The leaked minutes of the secret May 2004 LETAG meeting with leaders of the fossil-fuel industries provide an insight into the Howard Government's real approach to climate change policy. They illustrate starkly how the history of climate change policy in Australia has been a history of lying. The minutes show that while the Government pretends in public to take an even-handed approach, it has in fact been acting for the fossil-fuel industries and against their low-emissions competitors. While it tells the Australian public that it takes climate change seriously, in fact its political strategy has been to do just enough to prevent too many negative stories appearing in the media. While it claims to be acting, the Government only takes a reluctant step if it believes it is being outflanked by the Labor Party or losing too much face internationally. While it says it is responding to the scientific evidence, by gagging scientists it has tried to keep the public in the dark. As we will see, the Howard Government knows that voluntary programs with industry have had little or no effect on greenhouse gas emissions, yet it persists with failing programs because they provide the semblance of action. Perhaps the biggest lie of all is this: while the Howard Government claims to be concerned about climate change on behalf of all Australians, the truth revealed by its own words and actions is that it wants only to protect export revenue and the interests of a handful of corporations.

In summary, we have a Government that has allowed policy to be determined, even written, by the large corporations that have the most to lose from change. While this book is a history of the politics of climate change in Australia, it can also profitably be read as a case history of political deception and dishonesty. In what follows, I trace the Howard Government's efforts to undermine action on global warming, and the deceitfulness that has accompanied this – the dirty politics of global warming. Before doing so, it will help to outline the principles underlying the climate change debate and the structure of Australia's greenhouse gas emissions.

dioxide emissions of coal-fired power plants, are penetrating the market. Renewable energies, including photovoltaic, geothermal and tidal power, anticipate an investment boom. Wind power has already taken off and is an established industry in Europe. The era of the national grid may be ending, with energy supply likely to become more localised. It is no longer science fiction to foresee commercial buildings and even individual homes *supplying* electricity in addition to consuming it.

In short, corporations with a strategic view are now working hard to diversify so that they can survive and prosper in a world beyond fossil fuels. They are factoring in to their business plans the implications of climate change and anticipated climate change policies. The message has sunk in. In a world where public concern is reaching critical levels, this makes good business sense.

For 11 years the Howard Government has been the captive of the fossil-fuel lobby, and has been lagging well behind the thinking of progressive business. The Government's approach has always been a short-term one, driven by the demands of the traditional energy industries. A more strategic approach would have acknowledged that much more demanding greenhouse targets are inevitable, and that all industrialised and developing countries will be faced with the need to turn decisively to alternative energy sources. If that is accepted, a very different picture of Australia's 'competitive advantage' emerges. Instead of being locked into a world view in which Australia's future is seen to be built on digging up and burning or exporting the country's huge coal reserves, we could exploit our massive resources of alternatives, including solar, wind, tidal and geothermal energy. We could extend the use of natural gas and biomass fuels. Australia is very well placed in each of these areas compared with almost all European and industrialising Asian countries.

Australia's advantage in many alternative sources of energy is inherent, deriving from our large land area, sunny climate and long

Framing the debate / 19

coastline. And there is another resource that is not being exploited – our expertise. Australia could be a world leader in some of these new energy technologies, and in so doing provide thousands of jobs that are more skilled and sustainable, and better paid, than those provided by mining, shipping and burning fossil fuels.

In sum, before more demanding greenhouse gas reduction targets are introduced – a situation that is unavoidable in a decade or so – Australia has the opportunity to develop a strong advantage in energy efficiency and most alternative energy sources. When it comes to cutting greenhouse gases, rather than being at a unique *disadvantage* compared with other industrialised and developing countries, as the Government claims, Australia is particularly well placed for the longer term global energy revolution. Exploiting this advantage means seizing the opportunity at an early date to build on Australian expertise already acquired in areas such as energy efficiency and solar technology, and, more generally, moving early to reduce the long-term costs of conversion to non-fossil-fuel sources.

However, as this book will document, the Australian Government's position on climate change has taken us in the opposite direction, emphasising short-term economic goals and entrenching fossil fuels in the nation's energy economy. It has undermined research and commercial initiatives leading to a low-carbon future. Going back as far as Federation, it is hard to think of a political and policy failure that has been so damaging to the long-term national interest and so reckless in its disregard for the welfare of future generations of Australians.

THE ROLE OF CLIMATE SCIENCE

Anyone who participates in the greenhouse debate must take a stance on the science of climate change. For those who are not climate scientists, the decision is not what to believe but whom. The most important documents in the climate change debate are the five-yearly assessment reports produced by the Intergovernmental Panel on Climate Change (IPCC). The IPCC brings together 2500 of the world's climate scientists either as lead authors of various parts of its reports or as reviewers who read and comment on drafts. The IPCC is acknowledged by almost all governments (even Australia's) as the oracle on climate change. After the scientists have finalised the detailed reports, intense negotiations begin over the crucial 'summary for policy makers', with fossil-fuel lobbyists trying to ensure that uncertainties are accentuated and conclusions watered down.

Until recently there has been a widespread view, including among journalists and politicians, that the scientists can't agree. This has led many to take the comforting view that we need not worry too much about climate change until the scientists sort out their differences. As we will see later, the waters have also been muddied by a deliberate campaign by the fossil-fuel lobby and right-wing political groups to throw doubt on the conclusions of climate science. First in the United States, and then in Australia, a number of so-called climate sceptics have ventured highly critical opinions of the work of the IPCC and other climate scientists. In Australia, there are only four such sceptics with anything resembling scientific credentials – Bob Carter, William Kininmonth, Ian Plimer and Garth Paltridge. Only Carter and Paltridge can reasonably claim to be climate scientists with recognised stature in research.

In any field of science, genuine scepticism is an essential part of the progress of knowledge, as even the most widely accepted truths need to be challenged by contrary evidence and argument. However, the former CSIRO climate scientist Barrie Pittock argues that most of those who pose as scientific sceptics are better described as 'contrarians', because they are not interested in weighing the balance of evidence, but rather approach one position with extreme scepticism while failing to question the opposite view.[1]

In Chapter 10 I will argue that anti-greenhouse forces have exploited the desire of newspaper editors and journalists to appear to be balanced. In effect, the consensus view, based on studies published in peer-reviewed journals, is 'balanced' by the opinions of contrarians. Pittock characterises this as 'giving equal space to unequal scientific arguments'. Peer review is the means by which the science profession assesses the worth of all new work. Although it is not infallible, it is a generally effective filtering device, as most papers submitted to scientific journals (especially the best journals) are rejected and those that do get through the net often have to be revised before publication. The views of contrarian scientists have, virtually without exception, not been subjected to this process of professional review.

Given the profound importance of the topic, a genuine evidence-based challenge to the consensus view would make any climate scientist world-famous (and, if confirmed by other studies, would elicit the greatest sigh of relief in history). Yet in the two decades of debate over climate science, the contrarians have never been able to develop a coherent body of evidence that challenges the consensus view. Instead they have leapt upon individual pieces of evidence that appear to undermine the overall picture. One of the most extensively quoted of these was the satellite measurements of atmospheric temperatures which appeared to contradict the evidence of surface warming. In response to the doubts raised by sceptics, these satellite measurements were later shown to be wrong. Also heavily referenced was the so-called 'hockey stick' debate. A study by climate scientists combined various sets of data to map average temperatures in the northern hemisphere over the last 1000 years; it showed sharp increases in temperature in the twentieth century. This study was criticised for understating the degree of variation in temperature in earlier centuries. The critics claimed that large variations in the past probably mean that the twentieth century was not the warmest of the millennium. Further analysis by a number of scientists showed that the original paper did indeed underestimate the degree of variability over time, but it did not alter the conclusion that there was a sharp increase in temperature in the twentieth century and that it was the warmest century of the last ten.[2]

The climate change denialists in the media – especially columnists in the Murdoch press – seize on these studies, usually after they have been 'spun' and publicised by anti-greenhouse groups in the United States. When the claims are disproved, they simply move on to the next piece of apparently contrary evidence.

THE SCIENTIFIC CONSENSUS

Given the difficulty of sifting and synthesising the work of thousands of scientists in an area of knowledge that is extremely complex and undergoing rapid change, the IPCC process has been remarkable for its thoroughness. In 1996 the panel's Second Assessment Report galvanised the world into taking action. In some oft-quoted words, it concluded: 'the balance of evidence suggests a discernible human influence on global climate'. The state of climate science was summarised by Robert Watson, chair of the IPCC, in November 2000. Watson wrote:

> The overwhelming majority of scientific experts, whilst recognizing that scientific uncertainties exist, nonetheless believe that human-induced climate change is inevitable. Indeed, during the last few years, many parts of the world have suffered major heat waves, floods, droughts, fires and extreme weather events leading to significant economic losses and loss of life ... The question is not whether climate will change in response to human activities, but rather how much ... how fast ... and where.[3]

Watson (who was subsequently forced out of his position by the Bush Administration) went on to point out that the Earth's average surface temperature during the twentieth century was the warmest in over 1000 years, with the last two decades being the hottest of the century. The six warmest years of the twentieth century all occurred in the 1990s, with 1998 being the hottest. (Since Watson wrote this, we now know that 2005 was the hottest year on record.) The Third Assessment Report, published in 2001, reinforced these findings. It differed from the second report in two critical respects: the extent of expected warming had almost doubled; and the conclusions were drawn with a much greater degree of certainty. In addition, it found that rainfall patterns around the world are changing, the sea level is rising, glaciers are retreating, Arctic sea ice is thinning and the frequency of extreme weather events is increasing. The best estimate of sea-level rise is close to two millimetres per year, totalling ten centimetres since 1950. Tens of millions of people are at risk with land losses projected to range between 1 per cent in Egypt, 6 per cent in the Netherlands, 17.5 per cent in Bangladesh and 80 per cent in the Marshall Islands. Climate change is also expected to intensify the El Niño phenomenon, resulting in more severe droughts in Australia and more frequent and intense bushfires.

These changes in the Earth's climate are due to increased concentrations of greenhouse gases in the atmosphere arising from combustion of fossil fuels, deforestation and agricultural practices since the start of the industrial revolution in the eighteenth century. The Third Assessment Report in 2001 indicated that global mean surface temperatures will increase by between 1.4°C and 5.8°C by the year 2100[4], an increase on the estimated range of 1°C to 3.5°C in the IPCC's Second Assessment Report. (The wide range of the warming estimates is due not so much to uncertainties in the climate models but to doubts about how much carbon dioxide and other greenhouse gases will be pumped into the atmosphere in the coming decades.) This rate of warming is faster than anything the Earth has experienced for at least 10,000 years.

In 2001 it was generally believed that, with concerted action, we could stabilise concentrations of greenhouse gases in the atmosphere at double their pre-industrial levels of 280 parts per million. This now seems an unreachable target without rapid and deep cuts in global emissions. The Fourth Assessment Report of Working Group I of the IPCC was released in February 2007.[5] With the heightened awareness of climate change around the world, it attracted enormous media coverage. Although couched in the cautious language of consensus science, it sent a message too powerful to ignore. Changing climatic patterns previously judged 'likely' had become 'very likely', that is, with more than a 90 per cent chance of occurring. It stated that warming is now 'unequivocal' and noted that 11 of the previous 12 years rank among the 12 hottest years ever recorded. Warming in the Arctic is occurring at almost twice the global average, with the Arctic permafrost layer heating by up to 3°C in the last 20 years.

The Fourth Assessment Report warned that if greenhouse gases continue to be emitted at or above current rates, the consequences will be far-reaching. In perhaps the most likely scenario, the globe will warm by 3°C or more and sea-level rise will be in the range of 0.35 to 0.43 metres by the end of the century.† The IPCC projects sharp reductions in snow cover, the disappearance of Arctic sea ice in summer and more frequent heatwaves and floods. Ominously, it noted that due to uncertainty about their magnitude, positive feedback effects in the climate system were excluded from consideration in the report. These feedback effects could see the

---

† Some of the projections of future social and economic changes used by the IPCC are far-fetched and can be discounted. Others describe realistic possibilities. The most likely scenarios without concerted action are those called A1B and A1F1.

Earth warming much more quickly and with almost unimaginable consequences.

Warming is already giving rise to more extreme weather events. A heatwave in Europe during the summer of 2003 saw maximum temperatures reach 5°C above the long-term average. The French health ministry reported 14,802 additional deaths in August, mostly of old people unable to cope with sweltering temperatures peaking over three days.[6] Thousands more deaths than expected occurred in Germany, Spain and the United Kingdom. Under expected warming scenarios, disasters like this will become more common, with one study suggesting that the chance of heatwaves like that of 2003 could be 100 times higher over the next 40 years.[7]

Poor people in developing countries will suffer most from the effects of climate change. Droughts are expected to become more frequent and prolonged, especially in central Asia, Africa and the Middle East, leading to crop failures and famine. Ecosystems will struggle to keep up with shifting climates, with forests especially vulnerable. Large populations, mainly in tropical and sub-tropical regions, will be more at risk from diseases such as malaria, schistosomiasis (also known as bilharzia or snail fever) and dengue fever as the ranges for these diseases spread with shifting weather patterns. And while many regions will experience an overall drying, when the rains do come they are more likely to come as a deluge, so that floods are also expected to increase in frequency and magnitude in most regions of the world.

A great deal of climate change is already inevitable based on past greenhouse gas emissions. If consumption of fossil fuels were halved tomorrow, climate change would not be prevented. Such change has great inertia because most greenhouse gases persist in the atmosphere for at least a century before they can be absorbed by the oceans and the earth. Even if atmospheric concentrations of greenhouse gases were stabilised by 2050 – something that would require cutting current global emissions by up to 90 per cent – temperatures would continue to increase over several decades, and sea levels would continue to rise for hundreds of years.

This makes climate change unique among environmental problems. For most of these, the effect follows directly from the cause. If industrial pollution is causing smog that makes us sick, we can pass laws to reduce air pollution and smog will soon be reduced. Yet after a century or more of industrial activity, the effects of climate change are only now beginning to be felt; and the effect of what we do today will be felt in 50 or 100 years' time. This fact has allowed political leaders to defer action. If we had to live with the consequences of our own actions, we would have fixed the problem some time ago. Our response to climate change is thus a test of how much we care about future generations. Many people are moved to take action because they can imagine the lives of their grandchildren. Others don't seem to care.

Yet the future has arrived. The former UK secretary of state for environment Margaret Beckett observed that, 'We've known for some time that we have to worry about the impacts of climate change on our children's and grandchildren's generations. But we now have to worry about ourselves as well'.

Australia's Commonwealth Scientific and Industrial Research Organisation (CSIRO) has become one of the world's leading research agencies examining climate change. Much of its work focuses on the expected impacts on this country. It projects that average temperatures in capital cities will rise by as much as 3.4°C by 2070, meaning that we can expect a doubling or even a quadrupling of the number of days each year with temperatures over 35°C.[8] Rising sea levels will lead to higher storm surges, more frequent coastal flooding and damage to coastal ecosystems, including beaches being washed away. A substantial reduction in winter snow cover is already evident, affecting not only skiers but also the survival of the plants and animals that live in alpine environments. Fire researchers are anticipating more days of high and extreme fire danger over much of the continent in coming decades. In northern Australia we can expect permanent damage to the Great Barrier Reef within 30 years, due to bleaching as seas heat up. Warming is expected to lead to a higher incidence of diseases such as malaria, encephalitis, Ross River and dengue fever, with all of these spreading southwards.[9]

Such warnings cannot be dismissed as green scaremongering. These projections – backed by the UK Meteorological Office, the IPCC, the CSIRO, NASA and a dozen other scientific organisations around the world – are not about what might happen if we do not act now. They are the most likely outcomes even if the 1997 Kyoto Protocol, designed to limit the growth of greenhouse gas emissions, comes into force and does what it is supposed to do in the first commitment period. In other words, we cannot prevent climate change: it is rather a question of whether the nations of the world can agree on ways to prevent its effects being much worse.

WHY GLOBAL WARMING?

What are the essential processes of climate change? The explanation that follows is simplified, but I hope that it will serve to illustrate the causes of global warming and draw out some of the key political questions.

The presence of carbon dioxide (and certain other gases) in the atmosphere keeps the Earth some 30°C warmer than it would otherwise be. By horticultural analogy, this is called the 'greenhouse effect' and the heat-trapping gases are called 'greenhouse gases'. When fossil fuels are burnt, their carbon oxidises (combines with oxygen) to make carbon dioxide, a gas which escapes into the atmosphere. Carbon dioxide is also released from sources other than combustion of fossil fuels, notably deforestation.

There are two other important greenhouse gases: nitrous oxide arises mainly from the use of fertilisers, and methane is produced by livestock and from coalmines and waste dumps. But carbon dioxide produced by burning fossil fuels is the chief culprit in global warming.

Carbon occurs naturally in various forms in the atmosphere, the terrestrial biosphere (vegetation and micro-organisms in the soil), the oceans and deposits under the ground. Cycling of carbon between these media evolved over millennia into an equilibrium. But around 200 years ago, humans began to dig up coal in sizeable quantities, and about a century ago, large amounts of oil began to be extracted. The concentration of carbon dioxide in the atmosphere averaged 280 parts per million by volume in the pre-industrial era; it is now around 360 parts per million and on present trends is expected to exceed 500 parts per million by 2050. The concentration of carbon dioxide in the atmosphere will treble later in the century without determined action to cut global emissions. It is this increase in heat-trapping gases that enhances the natural greenhouse effect, resulting in global warming and associated changes in the Earth's climate.

Most of the coal deposits we extract today were laid down in the Carboniferous period, 360–280 million years ago, when successive layers of sediment formed rock covering thick layers of plant matter from lush tropical forests. Under pressure, and subject to rising temperatures the deeper they were pushed, the organic materials metamorphosed over very long periods into coal. Burial at a depth of over five kilometres is required to make the best quality coal. Brown coal (or lignite) is of lower quality; that is, its energy content is lower because it is wetter and less coalified; it therefore emits more carbon dioxide for each unit of usable energy generated. While lignites are around 60 per cent carbon, anthracites (or black coal) are over 92 per cent carbon. Coal dug up in Victoria in open-cut mines is brown, while that from underground mines in New South Wales and Queensland is black. Victorian electricity is dirtier but cheaper, while higher quality coal from the other two states can be exported as well as burnt here.

While the science of climate change is vastly more complicated than this, my outline is intended to emphasise one basic fact: recent dramatic climate change has come about because humans have been extracting carbon that has for millennia been safely stored in highly concentrated form, and converting it into carbon dioxide that ends up in the atmosphere. We have been disturbing the great carbon cycle so that there has been a net transfer of carbon from under the ground into the atmosphere. In other words, the genie is out of the bottle. Undoing the enhanced greenhouse effect will require returning the Earth's atmosphere to something approximating its previous equilibrium, and this will require getting increased carbon dioxide and other greenhouse gases out of the atmosphere and back into safe long-term storage, a process that will take centuries.

Three political points follow from this outline.

First, human-induced climate change is caused by an increasing *concentration* of greenhouse gases in the atmosphere. Annual flows add to stocks, so the rising concentration is the cumulative result of emissions over the last two centuries, the period in which humans have been burning fossil fuels in quantity. Therefore, when assessing the responsibility of nations for climate change, it is their contribution to the overall concentration that matters, not their current annual emissions. This gives the lie to US and Australian claims that the Kyoto Protocol is 'flawed' because developing countries are not required to cut their emissions, yet will account for more than half of annual emissions by around 2020. Industrialised countries are responsible for around 75 per cent of the increased greenhouse gases in the atmosphere today and it will be many decades beyond 2020 before developing countries account for half of the increased concentration of greenhouse gases. This is one reason why developing countries – and fair-minded people in rich countries – find the demands of the anti-Kyoto countries profoundly unjust.

Secondly, global warming is inherently a global problem. It is the *burning* rather than the *mining* of fossil fuels that causes the damage. Because of this, responsibility should be borne by the nation where combustion occurs. But when fossil fuels are burnt and greenhouse gases are released into the atmosphere, at present no-one has responsibility for the effects. Unlike urban air pollution, where the damage is felt close to the point of emission, the damage from a tonne of greenhouse gases emitted in Sydney is felt equally in Ankara and Abu Dhabi. In economic terms, the atmosphere is a 'common property resource' which is treated as a free waste dump. The problem can be tackled only by international agreement.

Thirdly, it is generally believed that reserves of conventional oil will last only another 30 to 40 years, while known reserves of coal will last for 200 to 300 years. If all of the known and hypothesised deposits of coal and oil were extracted and burnt, the concentration of carbon dioxide in the atmosphere would increase six-fold. As the Kyoto Protocol is designed to begin a process that will prevent a doubling or, at worst, a trebling of concentrations, it is clear that, at some point, a decision must be made to leave large amounts of exploitable carbon in the ground. Many people find it hard to accept that in order to prevent potentially catastrophic climate change we must stop burning fossil fuels. But this should not cause too much alarm; after all, we leave some exploitable reserves of uranium in the ground because we have serious reservations about the dangers of its use. And we do not clear every forest simply because we can.

PRINCIPLES OF JUSTICE

Deep ethical undercurrents swirl beneath international climate change negotiations. The principal questions are distributive: who is responsible for the problem, who will suffer most from climate change, and who will bear the costs of abatement measures? Nations and interest groups argue vociferously that certain proposals to tackle climate change are 'unfair'. In the lead-up to the Kyoto conference, the Australian Government argued that the European proposal for equal percentage cuts was unfair because it would impose a greater economic cost on Australia. Arab oil-exporting countries claim that any treaty requiring cuts in consumption of petroleum would be unfair to them and they have repeatedly demanded financial compensation for the impacts of the treaty (demands that have met with little sympathy). The United States, backed by Australia, has claimed that it would be unfair and ineffective to ask it to cut emissions when large developing countries such as India and China are not required to cut theirs.

The Alliance of Small Island States (AOSIS) has a particularly strong moral case. Some small island states are threatened by inundation due to rising sea levels, so their very survival is at stake. The group of developing countries known as G77 has insisted that it is unfair to ask developing countries to begin to cut their emissions when they have not been responsible for climate change, yet will suffer disproportionately from its effects.

There is also the consideration that reducing emissions will entail structural changes to economies, which means there will be winners and losers. For example, carbon taxes will disadvantage industries dependent on coal and oil, but will probably benefit other industries.

Each claim of unfairness is based on an implicit principle of justice or fairness, but these principles are rarely articulated. Yet there are two ethical principles that underlie environmental debates. The most important and widely accepted of these is 'polluter pays', in which the entity responsible for the pollution should be responsible for cleaning it up. Those countries most responsible for greenhouse gas emissions should be required to do most to reduce them. In practice, this is interpreted to mean that nations with high per capita emissions have an obligation to accept more stringent emission reduction targets.

The polluter-pays principle also implies that the polluter should compensate the victims. This idea is embedded in the United Nations Framework Convention on Climate Change (UNFCCC), the mother convention to the Kyoto Protocol, in the provisions that call on developed countries to provide financial support to developing countries for adaptation measures. The call by the United States and Australia for developing countries to be brought into the emission-cutting process sooner rather than later should be seen as an inversion of this principle, one that might be called 'victim pays'.

The second principle of fairness is that of 'ability to pay', under which, other things being equal, nations or other parties that are in a better position to pay should be required to do so. The principle that the wealthy should be required to contribute more is the foundation of progressive income taxation. If two nations had identical levels of emissions yet one were much richer, then fairness would demand that the richer country do more; hence the sense of moral outrage felt by many at demands that developing countries should be asked to adopt targets at the same time as rich countries.

Mention should also be made of the idea of rights, as these too enter the debate at many points, notably in discussion of emissions trading schemes. Prior to the Kyoto Protocol, each country had the privilege, arising from its sovereignty, to emit greenhouse gases into the atmosphere without limit. As our understanding of the enhanced greenhouse effect evolved, the morality of unrestricted emissions was challenged. The Kyoto Protocol is designed to place restrictions on the legal right to pollute, restrictions defined by targets for each country.

One further point can be made to illustrate the political power of the principles of justice outlined above. It is often argued by the Howard Government that since Australia is responsible for only around 1.4 per cent of the world's annual greenhouse gas emissions, we should not worry too much about reducing them, as any action of ours would have a minimal effect on global climate change. This 'pragmatic' argument contravenes the principles of both polluter pays and ability to pay, as Australia has very high emissions per person and is a wealthy nation. It is not only fallacious but dangerous in its implications, for it actually fails the test of pragmatism. If the world were made up of only 71 nations, each of which was responsible for 1.4 per cent of global emissions, then no-one would take any action and we would continue to emit greenhouse gases until catastrophe befell us.

More importantly, the argument undermines action by others. In 2000 Australia's then richest man, Kerry Packer, was the subject of widespread public criticism because of his tax minimisation schemes. The general view was expressed in parliament by Mark Latham, who linked the Tax Office's pursuit of Mr Packer to the media mogul's recent gambling spree in Las Vegas:

> If someone has enough money to blow $34 million at a casino, then they have enough money to pay more tax and help to build a better society. If someone has enough money to engage in conspicuous consumption, then they have a conspicuous responsibility to assist the most disadvantaged parts of society.[10]

In the press and on the airwaves Mr Packer was accused of shirking his responsibilities and undermining the integrity of the tax system. The public legitimately asked: if Australia's richest man does not pay his taxes, then why should we? His refusal to pay his fair share was seen as self-interested and immoral. This is precisely how Australia is seen by the rest of the world for its stance on climate change.

With all of the attention given to the different stances of countries in the North and South, one question of fairness has received almost no attention at all – the distribution within countries of the *effects* of climate change. While the Australian Government has often warned about the effect of reducing greenhouse gas emissions on workers in coal-dependent industries, no-one talks about the costs of *failing* to reduce emissions. For example, in some areas the costs to agricultural production are likely to be large, with changes to water resources, increased frequency of drought and extreme weather events such as hail-storms to contend with.

Very little can be said with certainty about the distribution of such costs within Australia, as almost no useful information is available (itself a serious failure on the part of the Federal Government). Some changes will have direct economic implications, while others will affect the levels of health or comfort experienced by people in their daily lives. The financial costs of mitigation measures will be met by private households (for instance, moving from newly flood-prone areas or installing air conditioners) and by the public sector (building dykes and repairing roads). In addition, there will be 'amenity costs' for people (such as living in a less pleasant climate, and loss of biodiversity), which are simply borne because individuals or governments are unable or unwilling to take measures to reduce their effects.

For Australian households, the principal costs are likely to involve protective measures such as 'climate-proofing' homes, insurance costs and medical expenses. Insurance costs are likely to increase substantially; indeed, it may be difficult to obtain insurance against cyclones and storms in some areas. This is already happening in the United Kingdom in response to the floods of November 2000, with a resultant decline in property prices in affected areas. Similarly, after devastating cyclones in Florida, insurance companies have been reluctant to provide cover.

In general, the effects of environmental degradation fall more heavily on the poor. Wealthier people can afford to live in areas with low air pollution, better water quality and superior amenities. In cases of local environmental decline, they are also better able to protect themselves through the exercise of political power. In the United States, air quality is worse in areas where low-income households are located. An Australian study of lead levels in children showed that children from families with low annual incomes have substantially higher levels of lead in their blood than children from families with higher incomes.[11] In Australia it is almost certain that those who will suffer most from the effects of climate change will be rural and Indigenous communities.

In the next chapter I will outline the nature and extent of Australia's greenhouse gas emissions. Armed with the facts, we will then be in a better position to assess some of the arguments used by the Government to evade taking action to cut our emissions.

3. AUSTRALIA'S EMISSIONS

DISPELLING THE MYTHS

Myths and trickery distort the discussion of Australia's greenhouse gas emissions. The Government and the fossil-fuel lobby have made many misleading claims based on these myths, so it is best to be clear about the numbers. The basic source of information is the annual inventory prepared by the AGO, the federal agency established after Kyoto to develop and implement climate change policy.†
A range of experts in the sectors responsible for emissions – energy, agriculture, waste, land-use change and forestry – assist with the inventory's preparation.

Throughout the industrialised world the bulk of greenhouse gas emissions arises from the burning of fossil fuels – coal, oil and gas. Methane emissions from agriculture and waste are also significant. Australia is the only industrialised country in which the clearing of land for agriculture results in substantial emissions. As we will see, this fact is responsible for a crucial loophole in the Kyoto Protocol inserted at the insistence of the Australian Government. It is also the basis for the dubious claim that Australia is 'on track' to meet its Kyoto target. For the moment, though, we will exclude land-clearing from the discussion so that we can see the overall trend.

The growth in Australia's greenhouse gas emissions from 1990

---

† Each inventory contains data on emissions two years previously, so the inventory published in 2006 is for the year 2004.

FIGURE 1: Growth in Australia's greenhouse gas emissions 1990–2004 (Mt CO_2-e) (excluding land-use change and forestry)

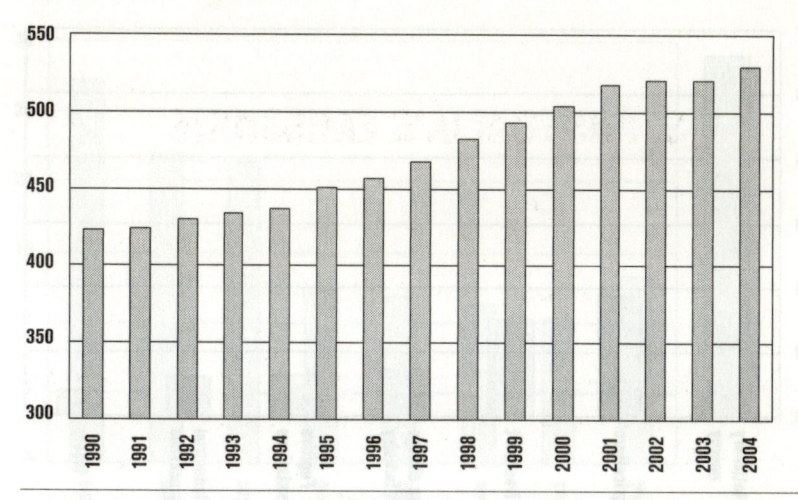

to 2004 is shown in Figure 1. The year 1990 is the base year for calculating emission targets under the Kyoto Protocol. In 1997, Australia agreed at Kyoto to limit emissions to an average 108 per cent of 1990 emissions across the five-year period from 2008 to 2012. Emissions are measured in millions of tonnes of carbon dioxide equivalent (Mt CO_2-e). Carbon dioxide, the main greenhouse gas, accounted for 74 per cent of all Australia's emissions in 2004, methane for 21 per cent, and nitrous oxide for 4 per cent. The latter two are converted into 'carbon dioxide equivalents' according to their global warming potential.

It is clear from Figure 1 that Australia's greenhouse gas emissions have risen rapidly since 1990, and since the Kyoto Protocol was agreed in 1997. From 1990 to 2004, excluding land-use change and forestry, total emissions in Australia increased by 25 per cent, a figure that may be compared with the 8 per cent increase assigned to Australia under the Kyoto Protocol and the 8 per cent *reduction* required over the same period for the European Union. The claim by the Government that its policies are working is manifestly untrue. In 2004, reliance on coal for electricity generation, the poor fuel efficiency of our vehicles, continued land-clearing and relatively high levels of emissions from agriculture. Even so, and contrary to the Government's claims, Australia's emissions profile is not markedly different from other developed countries. Our industry does a little worse, and our residential and commercial sectors slightly better compared with the average emissions of countries in the Organisation for Economic Co-operation and Development (OECD).[3] Our per capita transport emissions are one-third higher than the OECD average, but substantially less than those in the United States and Canada. This figure goes some way to dispelling the belief that Australia's low population density necessitates proportionally higher transport use.

Australia is a large country, but around 62 per cent of all fuel used for land travel is consumed in urban areas.[4] The reality is that most Australians do not spend their time cruising across the wide brown land, but rather sit in traffic jams in the congested brown city. It is not so much geography as the lack of alternatives to road freight that distinguishes Australia from Europe. Our passenger cars are also particularly inefficient. In the late 1990s the average Australian car was getting about the same number of kilometres per litre as the average car in 1971.

Since our climate change debate has been dominated by concern over the potential effects on Australian industry, it's worth noting that one industry – aluminium smelting – accounts for an enormous proportion of Australia's industrial emissions, a fact that explains the strong interventions of the aluminium industry in the policy debate. Australia's six smelters account for around 16 per cent of total national electricity consumption (the emissions from which do not appear in the AGO's inventory as the responsibility of the aluminium sector). However, Australia's disproportionately large non-ferrous metals industry does not make us unique. Germany, Japan, the Netherlands and Norway all have levels of dependence on emission-intensive industries – iron and steel, chemicals and petrochemicals – comparable to Australia's. Russia and Canada are also heavy exporters of energy. Yet all of these countries have ratified the Kyoto Protocol.

While the Government makes sweeping claims about the emissions intensity of the Australian economy (that is, the emissions per unit of gross domestic product, or GDP) and thus its unique exposure to competition from abroad, the great bulk of our GDP is accounted for by goods and services that are not significantly affected by international trade. Think of housing, the government sector, land transport and a thousand service industries ranging from retailing to hairdressing to legal advice. For around three-quarters of the economy, trade is largely irrelevant.

Yet overseas competition is of concern to a number of industries. A company or industry may face a competitive disadvantage from policies that increase energy prices, such as a carbon tax, but only if three conditions apply: it must be emissions intensive, it must be exposed to import or export competition, and that competition must come from countries that do not have similar policies (that is, those not part of the Kyoto Protocol). The only Australian industries that meet these criteria are aluminium, alumina, steel, other non-ferrous metals (mainly nickel), liquefied natural gas and gold.[5] These industries account for around 1.5 per cent of Australia's GDP and 19 per cent of merchandise exports (those exports excluding services such as education and tourism).[6] If Australia were to ratify the Kyoto Protocol and impose policies that increased the price of energy, it would be a straightforward matter to provide tax offsets for exports whose competitiveness would be affected, just as the goods and services tax is not applied to exports. This small problem with a simple solution has nevertheless been driving the Howard Government's intransigence on Kyoto. It demonstrates how a powerful industry lobby can hold a country to ransom when it has captured a sympathetic government.

It is worth examining another argument the Prime Minister has used to bamboozle the public. 'Unlike most developed countries,' he has often said, 'Australia is a net exporter of energy and that puts us in a very special position'. Australia is easily the world's biggest coal exporter, accounting for around 30 per cent of global trade. In addition, coal is Australia's largest individual merchandise export, forming 16 per cent of the total.[7] Yet our energy exports have no bearing on Australia's emission reduction obligations at all. The emissions from our exports of coal, gas and oil are counted in the country where they are burnt. It is likely that, in response to Kyoto, *other* countries may decide to cut their imports of fossil fuels, but there is nothing Australia can do about that – except perhaps to try to sabotage the Kyoto Protocol.

The Government also claims that Australia's fossil-fuel dependence makes it harder for us to cut our emissions. In fact the opposite is more likely to be the case. The cost of reducing emissions by, say, 10 per cent depends not on how much fossil fuel is burnt but on how efficiently a country burns it. If fuel is used wastefully, it will cost less to reduce consumption. As an economy reduces its emissions, it will start with the cheapest abatement measures (energy savings) and then move to the more expensive measures by replacing energy-hungry equipment and switching from high-emission sources such as coal to low-emissions sources such as natural gas and renewables. Thus countries that have been reducing their reliance on fossil fuels for some time will probably have eliminated the least efficient uses of these fuels first. This was the case in Japan after the oil shocks in the 1970s and early 1980s, when oil prices doubled overnight. Similarly, countries that have built nuclear power plants have tended to replace the least efficient coal-fired plants.

It is sometimes said that in reducing emissions those responsible will first 'pick the low-hanging fruit'. If more fruit is wanted, more effort must be expended to get it from the higher branches.

Compared to most other OECD countries, Australia has not yet picked the low-hanging fruit. This is because fossil fuels in Australia have been cheap and abundant. That was the message of the OECD's International Energy Agency when it reviewed Australia's energy economy in the late 1990s.[8] It is also the message of the most comprehensive analysis of Australia's energy efficiency performance, carried out by the foremost expert in the area, Lee Schipper of the Lawrence Berkeley Laboratories.[9] The analysis concluded that, while the story varies from sector to sector, Australia's overall energy efficiency is poor compared with other OECD countries. (The Government refused to release Schipper's report.)

In Australia several studies have looked at cost-effective ways to reduce greenhouse gas emissions. These studies estimate that Australia could cut its energy consumption by at least 20 per cent at *zero cost* – enough to reach the Kyoto target. In most instances the payback period for the investment in energy efficiency equipment is only a year or two.[10]

Because Australia has so much low-hanging fruit, we could easily meet our Kyoto target and go well below it, giving us surplus emission permits to sell to countries like Japan and Spain, which will struggle to meet their targets. With the right domestic policies, ratifying the Kyoto Protocol could have been a lucrative opportunity for Australia. However, those who would gain from such a move – the public, the low-emissions industries and the environment itself – have always had less political power than those who would lose, particularly the all-powerful export coal lobby. As will become apparent in Chapter 6, the collapse in emissions from land-clearing has permitted the Howard Government to conceal from the public the sorry consequences of Australia's burgeoning greenhouse gas emissions. To get us to that point we now turn to the beginnings of the greenhouse story in Australia.

4. BOLD DECLARATIONS

EARLY HISTORY

In the course of researching this book, I came upon a remarkable confidential report prepared by the Office of National Assessments (ONA) entitled *Fossil fuels and the greenhouse effect*.[1] The most striking feature of this document is the date on it – November 1981. It turns out that the Federal Government had been alerted by its primary intelligence-gathering agency to the problem of climate change much earlier than anyone thought. The document, extraordinary for its prescience, begins with a simple and accurate statement of the problem and the likely consequences, an assessment that is essentially unchanged today. It canvasses the possibility of a doubling of atmospheric carbon dioxide by 2050 and a quadrupling by 2100. The ONA foresaw far-reaching economic and political effects, predicting that by the end of the twentieth century, concerns about greenhouse 'could culminate in pressure for action to restrict fossil fuel use'. It went on to observe: 'There are potentially adverse implications from these developments (if realised) for the security of Australia's export markets for coal beyond the end of the century'.

The ONA told the Government that considerable scientific uncertainty remained, but that even so the public was liable to become alarmed. 'The discussion is clouded by the attention of environmentalists, and by the partisans of nuclear power, who have their various reasons for discouraging the use of fossil fuels.' After noting that many countries planned to expand their use of fossil fuels, the ONA considered likely political responses.

> So far there is no anti-fossil fuel lobby comparable to the anti-nuclear groups, although some environmental groups are beginning to express concern. Perhaps because the problem is merely the gradual increase of a non-poisonous substance which has always been present, public alarm will only be generated by manifest change, or a threat of it, such as a rise in the sea level. Nevertheless, increasing awareness of the problem could begin to generate an opposition to fossil fuels, encouraged by pro-nuclear lobbies and environmental groups, in this decade.

In today's politicised intelligence agencies, it is hard to imagine the ONA providing such a balanced and well-informed analysis. The analysts who prepared the document showed astonishing insight, accurately predicting in 1981 the state of play in greenhouse policy 25 years later. For instance, they observed that the general opinion was that 'at present nuclear power is the only large-scale alternative to fossil fuels'.

Unsurprisingly, though, the report seems to have had no impact in Canberra. One senior bureaucrat scrawled on the document: 'This report is not middle of the road as I understand [the] situation to be'. As late as 1989, when the Government established a working group to examine the future of energy supply in Australia, the question of global warming was not part of its brief and had to be added as an afterthought. However, Prime Minister Hawke appears to have recognised that it was a looming problem that would not go away. In a media release dated 6 April 1989, he declared that 'greenhouse cannot be dismissed as just another environmental issue' (itself a revealing comment on the still-marginal status of 'the environment'). 'It has the potential to change fundamentally within a single lifetime the way all nations and peoples live and work. It clearly signals that we must reassess the way in which we use the earth's resources.'[2] In the same year, Hawke also warned that the threats posed by climate change could lead to a reassessment of Australians' opposition to nuclear power.[3]

Curiously, the 1981 ONA report was addressed to 'Mr J. Howard, Department of National Development and Energy'. In 1981 the namesake of that obscure public servant, the man who is now our Prime Minister, was Treasurer in the Fraser Government and it is quite possible that he read the ONA report at the time. If so, it is now clear from his actions that he was wholly unpersuaded by it. On the other hand, the script for the Prime Minister's strong advocacy of nuclear power in 2006 could have been taken directly from the 1981 document.

THE TORONTO TARGET

Climate change emerged as a public issue in Australia in 1988, and the most notable feature of those early years was the progressive position taken in the world community by the Australian Government. The Hawke Labor Government proved willing to adopt, in principle at least, ambitious targets for reducing emissions. In mid-1989 the governments of New South Wales, Victoria and Western Australia publicly adopted the Toronto Target – which aimed to reduce carbon dioxide emissions to 80 per cent of 1988 levels by 2005 – 'as an interim objective for planning purposes', even before there was any information on the means by which such a target could be met in Australia. State governments were spared the need to develop any actual plans by the issue rapidly moving into the national arena. In April 1989 the Hawke Government set up a National Greenhouse Advisory Committee of scientific advisers and a Prime Ministerial Working Group to assess achievable targets.

The greenhouse debate soon became polarised between the environment ministers and their department on the one hand, and the energy ministers and their department on the other. Climate change was seen as primarily an environmental issue, and the environment department urged action commensurate with the scale of the problem. However, the energy (and to a lesser extent, transport) departments are responsible for most of the means of reducing greenhouse gas emissions while also being the advocates within Government for the polluting industries.

In October 1990 the Commonwealth Government followed the states and also adopted the Toronto Target of a 20 per cent reduction, a target that in retrospect appears hopelessly optimistic. However, adoption of the target came with a vital caveat, one designed to appease industry and which presaged much that was to come. The Commonwealth made it clear that 'the Government will not proceed with measures which have net adverse economic impacts nationally or on Australia's trade competitiveness in the absence of similar action by major greenhouse gas producing countries'.

The next milestone was reached in June 1992. At the UN Conference on Environment and Development in Rio – the famous 'Earth Summit' – Australia was one of the 155 nations that signed the United Nations Framework Convention on Climate Change, the basic international instrument to deal with climate change.[4] Article 4.2 of the convention designated 2000 as the year by which signatories would endeavour to return greenhouse gas emissions to their 1990 levels. At this stage the Labor Government was still making bold declarations and trading on its high standing internationally over environmental issues.

WINDOW-DRESSING: THE NGRS

In December 1992 state and federal governments endorsed the National Greenhouse Response Strategy (NGRS). It was intended to be a comprehensive approach to reducing emissions, a collection of measures that, if carried out with vigour, would have made a small but significant impact on emissions growth. The plan was to pursue more efficient energy use and promote alternative fuels 'wherever economically efficient'.[5] A number of measures were listed, mainly concerning the efficiency of electricity and gas markets, requiring utilities to report performance against greenhouse indicators, and encouraging greater use of renewable energy.

However, these measures were systematically ignored. Nearly every major energy supply decision taken by state and territory governments in the following years favoured the options with the *higher* greenhouse gas emissions, including new coal-fired power stations and the extension of the electricity grid into areas previously served by remote area power systems and where renewable energy sources would have been cheaper. The electricity industry, through its organisation the Electricity Supply Association of Australia, has at all times resisted any serious measures to attempt to limit emissions from coal burning, although a few of its member companies have taken a more progressive position. At this stage of the debate, the industry did not feel the need to feign concern over climate change. Even so, the alarm bells started to ring inside the boardrooms of the big energy companies and they began to develop sophisticated strategies to stop any measures that would affect their commercial interests. In this, they were a decade ahead of the environment movement.

Some early skirmishes reinforced how high the stakes were. In a 1994 case before the NSW Land and Environment Court, approval of a new 135-megawatt coal-washery power plant in the Hunter

Valley was challenged on the grounds that there was no need for it, given the massive over-capacity of electricity generation in New South Wales, and that over its 30-year operating life it would increase carbon dioxide emissions by between 12 and 28 million tonnes compared with the alternatives. The court accepted that the power station 'will emit CO2, which is a greenhouse gas, and will contribute to the enhanced greenhouse effect, a matter of national and international concern'.[6] However, the court noted that there was no legal obligation on polluters to cut their emissions and could find no guidance as to the weight, if any, that should be given in decisions about actual projects.

Such cases demonstrated the radical disconnection in the minds of politicians and policy-makers between decisions over energy supply and the problem of climate change. It would be another decade before those making energy decisions automatically considered the greenhouse implications, if only to dismiss them – an object lesson in just how long it takes for people to change their way of thinking, even in the face of compelling evidence.

The National Greenhouse Response Strategy also proposed minimum performance standards for appliances, fuel efficiency improvements in new motor vehicles, provision of information for energy users, and more research and development.[7] The emphasis was on voluntary codes, which were never going to be sufficient to reduce emissions by the desired levels. Even these rather mild policy proposals ran into resistance from industry. The early 1990s were a period in which neo-liberal ideology, aimed at removing as many restrictions on business as possible, was at its peak. Indeed, in many instances the demands of ideology were more stringent than the demands of industry groups. Nevertheless, in this period, almost all large companies and industry associations in Australia opposed Government proposals to reduce greenhouse gas emissions.

A comprehensive review of the National Greenhouse Response Strategy in 1995 concluded that the strategy had 'failed to make any impact on Australia's greenhouse gas emissions. After two years of its operation, there is no evidence that even one tonne of carbon emissions has been saved as a result of the NGRS'.[8] After the failure of the NGRS, the Keating Government made a last attempt to retain a modicum of credibility on greenhouse – the 'Greenhouse 21C' package of March 1995 – but that too, being based on voluntary compliance, failed.

THE GREENHOUSE CHALLENGE PROGRAM

In 1995 the Labor Government used the threat of a carbon tax to persuade industry to accept a new initiative, the Greenhouse Challenge Program (GCP), a voluntary scheme enthusiastically taken up by the Howard Government after the 1996 election. Industry, though, had to be persuaded to participate, even in such an undemanding scheme. Within the federal bureaucracy there was a difference of view over whether green groups as well as industry ones should be invited to the first meeting, with one senior bureaucrat shouting: 'But they're the enemy!'[9]

Once it became clear that the Greenhouse Challenge Program would be an effective means of heading off mandatory measures, it did not take long for the fossil-fuel industries to back it eagerly. In a private letter to Prime Minister Howard in December 1997, the head of the main fossil-fuel lobby group, the Australian Industry Greenhouse Network, enthusiastically endorsed the program: 'It is also particularly heartening for the AIGN and its members who were instrumental in the establishment of the Challenge and continue to promote the program with wider industry at every opportunity'.[10]

The program involved voluntary agreements with major corporations responsible for the bulk of emissions from the industrial and energy sectors; it was subsequently extended to include medium and small enterprises. Like previous programs, the GCP is based on 'no regrets' measures, that is, measures that are worthwhile undertaking for their energy savings, irrespective of their impact on greenhouse gas reductions.

Soon after its inception, reluctant major emitting firms came under strong pressure from within the corporate world to sign up to voluntary agreements as an indication of good faith. There is no doubt that some of the firms that complied have made serious efforts to develop emissions reductions plans – Energy Australia, a NSW electricity distributor, is a case in point. In addition, the Greenhouse Challenge Program has helped to raise awareness within the business community about the significance of climate change and the need for the corporate sector to play a major role. However, the effectiveness of the GCP overall has to date been very limited. Its essential flaw is that many of the measures specified are those that the companies planned to undertake anyway, as part of normal commercial operations. In fact, although the details of the agreements remain secret, there are good grounds for believing that the emissions reductions claimed by the program were (and are) gross exaggerations, and that it will not significantly reduce Australia's emissions below the levels they would reach regardless.

Early in its history, the Commonwealth Government commissioned an independent study that attempted a proper evaluation of the effectiveness of the GCP.[11] This study was not made public until the former energy minister Senator Warwick Parer inadvertently agreed to release it under questioning from a Senate Estimates Committee in 1996.[12] The report, dated July 1996, was carried out by two leading energy consulting firms and included a detailed assessment of the first four confidential agreements signed under the program, those with BHP, Shell, CRA and ICI, some of the biggest emitters in Australia. The authors asked which of the emission-cutting projects that these companies proposed to undertake would have been implemented anyway under business-as-usual conditions. Their report concluded that 'about 83 per cent of the emissions reduction would most likely be realised in a BAU [business-as-usual] scenario'. In other words, only 17 per cent of the emission reductions planned by these companies could in any sense be attributed to the Greenhouse Challenge Program.

The Government has put a great deal of effort into promoting the signatory companies as responsible corporate citizens. Its promotional material to prospective clients stated that 'marketing of your environmental leadership will deliver significant, long-term public relations benefits … A sophisticated marketing campaign will highlight the achievements of industry'. Each batch of agreements has been launched with great fanfare, often by more than one minister, and expensive publications extolling the achievements of major firms are commonplace. The Government has bought full-page newspaper advertisements congratulating the major firms for their commitment. Such an advertisement in a major broadsheet costs at least $30,000.

The Greenhouse Challenge Program's marketing strategy was explicitly designed to advertise the green credentials of participating firms, some of which are honest enough to admit that part of their motivation for joining the challenge was to promote a 'clean and green' image.[13]

GREENPOWER

Voluntary schemes have also been directed at consumers. New South Wales was the first state to establish a Greenpower scheme, which was offered in 1996 by Energy Australia.[14] These schemes allow both households and business customers to opt to pay for the electricity they use to be generated from renewable sources such as wind power.

Since those households most concerned about greenhouse gas emissions are the target market, Greenpower schemes have been characterised by Richard Denniss as a 'tax on concern'.[15] Greenpower customers must be willing to pay a premium for their clean electricity and many ask, 'Why should I be asked to pay more for doing the right thing?'

These schemes suffer from most of the problems of voluntary approaches to environmental problems. Early market research indicated that around two-thirds of Australian households would be willing to pay more for electricity from 'green sources' (with a much lower share of business customers). Research for Origin Energy forecast in 1995 that by the year 2000, between 26 and 30 per cent of residential customers would be participating in such schemes.[16] However, by July 2006, less than 4 per cent of all households had signed up.[17] If new customers continue to join at these rates, it will be more than 20 years before Greenpower schemes make a serious dent in electricity emissions.

These discouraging figures should not be interpreted as showing a lack of concern about climate change on the part of Australian citizens, or an unwillingness to change. A change from an 'opt in' to an 'opt out' system would undoubtedly see a large increase in the number of participating households. More importantly, the survey results showing a widespread willingness to pay also imply that there would be popular acceptance if a government were to introduce compulsory measures, such as a carbon tax or mandated shifts to renewables. Climate change is far too big a problem to be left to the goodwill of individual citizens.

Despite the manifest failure of the voluntary approach it continues to be the basis of Government policy, for it serves the purpose of giving the impression that something is being done while imposing no constraint on the big polluters. But while Australian governments were working hard to avoid taking action, events abroad were on the march and it has been the international momentum to act that has always defined the terms of the greenhouse debate in this country.

5. THE ROAD TO KYOTO

SPECIAL PLEADING

The Kyoto conference in November 1997 marked a watershed not only for international efforts to combat climate change but also for Australia's reputation as a responsible global citizen. Our reputation was forfeited – and that was *before* the Howard Government repudiated the treaty it had agreed to at Kyoto. From having been a leading player in the 1992 Rio conference, by the end of 1997 Australia was being described in the world's press as a 'pariah nation'.

Throughout 1996 and 1997 the Howard Government mounted a vigorous and expensive international campaign to make its case. As the campaign evolved, it became increasingly clear that the intention of the Australian Government was to undermine the proposal for mandatory emission reductions, especially the model proposed by the European Union that would have required uniform reductions for industrialised countries of 15 per cent below 1990 levels by 2010.

The year 1997 was one of intense Australian diplomatic activity, with frequent ministerial journeys abroad. Officers of the Department of Foreign Affairs and Trade (DFAT) and the Australian Bureau of Agricultural and Resource Economics (ABARE) toured the globe attempting to persuade the rest of the world of the merits of Australia's case. Essentially, the Government argued that due to our heavy reliance on fossil fuels a uniform target would impose an unfair economic burden on Australia. It advocated a complicated formula for 'differentiated' targets, under which Australia would be assigned a more lenient target than other countries. Ministers claimed that such a proposal was consistent with the reference to 'common but differentiated responsibilities' in the UN Framework Convention on Climate Change.

Australia proposed a number of 'indicators' to determine the size of reduction targets, including growth in GDP, population growth, emissions intensity of the national and export economy, and dependence on fossil-fuel trade. It was of course no coincidence that under all of these indicators Australia would be assigned a more lenient target. The most glaring objection to the Australian differentiation argument was that it directly contradicted the polluter-pays principle. It was curious to observe economic policy advisers – noted for hard-nosed, 'face-the-consequences' prescriptions based on economic theory – abandon this principle in the case of climate change and discover an entirely novel 'equity' basis for their argument.

Australia's differentiation proposal was neither equitable nor workable, and it was met with dismay by other nations, especially developing ones. It was mired in contradiction. The Howard Government argued that developing countries should be brought into any agreement to reduce emissions because they will be responsible for most of the world's future emissions due to their high rates of population and economic growth. But when it came to Australia, the same high rates of population and economic growth were seen as a reason for assigning easier targets. Other governments and experts around the world took the view that Australia's heavy dependence on fossil fuels would make it *easier* for it to cut emissions, and that Australia's exceptionally high per capita emissions rendered perverse our pleas for special consideration.

Throughout 1997 various senior ministers returned from abroad claiming that the world was being won around to Australia's position. Environment minister Robert Hill was publicly upbeat. Deputy prime minister Tim Fischer returned from Paris in May claiming that Australia had made a breakthrough in its efforts to persuade other countries of the error of uniform targets. 'We're no longer alone as recently portrayed', he said.[1] Prime Minister Howard said in the aftermath of his visit to the United States in early July: 'I got a lot further on greenhouse gas emissions than I ever dreamt possible'.[2] However, it was clear to close observers that the reality was quite different. The Foreign Minister Alexander Downer inadvertently gave the game away in a speech to a Melbourne business seminar on 7 July by conceding that Australia's attempts to put its case were sometimes met with 'quite openly hostile opposition'.[3]

The Australian Government's case was subjected to a devastating attack from Timothy Wirth, US under-secretary for global affairs, during a Canberra–Washington satellite linkup in July 1997. Much of the Australian lobbying had been directed at the Clinton Administration, yet Wirth said that the US Government did not understand Australia's differentiation position: 'There's … an Australian suggestion that there be some sort of differentiation … We look forward to really learning what that means. We're not sure what differentiation means'. These comments were a sharp diplomatic slap in the face for the Prime Minister, since he had a week or so earlier met President Clinton to explain the Australian position. Howard's case was based on economic modelling by the Australian Bureau of Agricultural and Resource Economics. Asked about this modelling, Wirth replied that we should 'look at what those people are smoking'.[4]

At home and abroad, Australia was seen to be pursuing narrow self-interest with little regard for the environment or the diplomatic implications of demanding special concessions. Writing in June 1997 from New York, *Sydney Morning Herald* correspondent James Woodford observed:

Nothing so far has won the world over to Australia's cause and there is every indication that the world will not tolerate anything but acceptance of binding greenhouse gas targets ...

No Australian would have enjoyed joining the *Herald* or the ABC at the press conference held by three British Cabinet ministers, who effectively humiliated Australia in front of the world's media.

The British Foreign Secretary, Mr Robin Cook, was so sarcastic in his put-down of Australia's stance on greenhouse that almost the entire room burst into sniggers at the Federal Government's expense.[5]

At the same time, the UK's Prime Minister Tony Blair, in a barely disguised reference to Australia, demanded an end to 'special pleading' by industrialised countries.

The list of world leaders making clear that Australia's was an unacceptable path had become steadily longer. In November 1996, Clinton had given a speech in Port Douglas, Queensland, in which he criticised Australia's opposition to binding targets. In April 1997, Japanese prime minister Hashimoto said, while in Australia, that the Australian position would be hard to sell to the rest of the world. In May 1997, German chancellor Helmut Kohl, also on a visit to Australia, pointedly refused to acknowledge the Government's arguments. In June 1997, a spokesman for the UK's environment minister said: 'The Australian proposal flies in the face of the polluter-pays principle. As a high per capita emitter, Australia should be doing more – not less – than others if there were to be differentiation'.[6]

THE MINISTER FOR COAL

In the crucial period leading up to the Kyoto conference, Australia's minister for resources and energy was Senator Warwick Parer from Queensland, an untiring defender of the fossil-fuel industries in general and the coal industry in particular. In the 1970s Parer was a senior executive with Utah Mining, one of the largest coal producers in Australia, and in 1978 he became chair of the Australian Coal Exporters industry body. He entered the Senate in 1984 but remained involved in coalmining through his chairmanship of Queensland Coal Mine Management, a position from which he resigned only when he became a minister in March 1996.[7]

Throughout 1997 Parer issued media releases and gave speeches talking up the future of the Australian coal industry, lauding coal as the 'corner-stone of economic growth in the Asian region well into the next century' and praising 'clean coal'.[8] He made it clear that the Government would refuse to take any measures to reduce emissions that would, in his view, affect economic growth or employment.[9] So preoccupied was he with defending the coal industry that Parer seemed never fully to comprehend the issues involved in greenhouse science and policy. Indeed, in an address to an industry conference in March 1997 Parer actually declared that he did not believe in the greenhouse effect, thereby contradicting the official Government position: 'I don't have any figures to back this up, but I think people will say in ten years that it [greenhouse] was the Club of Rome'.[10] Ten years on, it is clear that he could not have been more wrong, yet Parer was an early influence on the Prime Minister's entrenched attitude to climate change.

In March 1998 Parer became embroiled in a scandal that was ultimately to lead to his downfall. It was revealed that he had breached the Prime Minister's ministerial guidelines by holding $2 million of shares in a company that owned three coalmines in Queensland.

The Prime Minister defended Senator Parer, described in the press as his 'one-time numbers man'.[11] For many years Howard and Parer had shared a flat in Canberra. Writing in *The Australian Financial Review*, Michelle Grattan expressed the common view of independent opinion: 'Senator Parer has been caught on toast with a potential conflict of interest. He has a big investment in a company operating in the coalmining sector, over which he has direct ministerial responsibility'.[12]

When the scandal broke, the Australian Democrats were particularly incensed at Parer's earlier decision to abolish the Energy Research and Development Corporation, a body whose purpose was to develop energy technologies that would provide a substitute for coal. Senator Lees said: 'How can a Government minister contribute to a decision which effectively nobbles the development of competition to an industry in which he has such a significant interest?'[13]

More scandals followed, and Parer hung on for a time with Howard's support, but he was severely wounded and quietly resigned from the ministry in October 1998 and the Senate in February 1999, after which he was appointed director of several coalmining companies.[14] The fact that Howard appointed as the minister for resources and energy a man who rejected greenhouse science, defended the interests of the coal industry at every opportunity and had a large personal financial stake in coalmining was symbolic of his approach to climate change.

THE ABARE EMBARRASSMENT

To support its case at home and abroad, the Government asked the Australian Bureau of Agricultural and Resource Economics (ABARE) to provide estimates of the costs of cutting emissions. Using its MEGABARE model of the economy, the bureau's results provided the basis for a number of publications, including two that were carried around the globe in the briefcases of ministers and public servants in their mission to influence world opinion.

From the outset ABARE's modelling came under severe criticism, both for what it included and for what it left out. In June 1997, a statement was signed by 131 professional economists – including 16 professors of economics – declaring that the bureau's modelling overstated the costs of abatement measures and underestimated the benefits. The economists said that 'policy options are available that would slow climate change without harming living standards in Australia, and these may in fact improve Australian productivity in the long term'.[15] Critics also pointed out, inter alia, that the MEGA-BARE model failed to allow for technological change, seriously overstated the likelihood of jobs going offshore and presented its estimates in a grossly misleading way.[16]

Ironically, careful examination of the modelling revealed that the costs of reducing emissions in Australia would be extremely small. The Government could claim that Australia faced ruinous costs only because ABARE used a number of statistical tricks.[17] For instance, the results indicated that real gross national expenditure would fall by 0.49 per cent in 2020 if the European proposal for uniform 15 per cent cuts were taken up. The point lost on most commentators, including Government ministers, was that this did not mean that the economic *growth rate* would be lower by this amount, but that absolute levels of real gross national expenditure would be lower. This is a very small change by any standard. A projected fall in gross national expenditure by half a per cent over a 25-year period would be swamped by many other changes in the economy. It was pointed out by economist John Quiggin, then at the Australian National University, that if the Australian economy were to grow by an annual average of 3.5 per cent, then per capita incomes would reach double their prevailing levels around 1 January 2025. If Australia reduced its emissions, according to the estimates, the doubling of per capita incomes would have to wait until around 1 March 2025, a delay of a mere two months.

On advice from ABARE, the minister for resources and energy, Senator Parer, declared in the Senate on 26 November 1996, and many times subsequently, that the cost of stabilising greenhouse gas emissions at 1990 levels by the year 2020 would 'be equivalent to a ... reduction in the savings of a family of four of about $7600'. The only way for ABARE to make its numbers 'look big' was to take a series of very small numbers over a very long period (25 years from 1996 to 2020), add them, and then calculate the impact on 'a family of four'. The $7600 per 'average family' should in truth have been compared to accumulated expenditure over the same 25-year period. In present value terms, that would be around $1.86 million. Exactly the same form of deception was used in 2006 by Government ministers claiming that ratification of the Kyoto Protocol would see wages fall by 20 per cent. It is an absurd and dishonest claim, yet it has been reproduced uncritically by the media.

The most damaging criticism of ABARE's modelling work emerged in May 1997. Parer revealed to the Senate that most of the funding for the research had been received from the fossil-fuel industry, including the Australian Coal Association, the Australian Aluminium Council, BHP, CRA, the Business Council of Australia, the Electricity Supply Association of Australia, Exxon, Mobil and Texaco.[18] These organisations paid $50,000 annually for a seat on the steering committee. The Australian Conservation Foundation (ACF) subsequently applied for membership and asked for the fee to be waived. The executive director of ABARE, Dr Brian Fisher, refused to do so, without providing an explanation.[19]

The question all except those involved asked themselves was what these companies expected to receive in return for $50,000 per year. In its promotional material, the bureau told prospective contributors that project management would be guided by members of the steering committee, who would provide 'a sounding board on policy, research and strategic issues'.[20] The Australian Aluminium Council (AAC) – probably the most powerful industry group in the greenhouse debate – was enthusiastic. In October 1997, its executive committee agreed to 'provide strong support' to ABARE's modelling and allocated $50,000 from its budget.[21] It is evident that these corporations and business associations would not have continued funding the modelling work if the results were not serving their commercial interests. It would be naive, if not foolish, to believe otherwise.

The ethical difficulty with industry funding was not lost on Professor Alan Powell of Monash University, an eminent economist and one of Australia's foremost economic modellers. Professor Powell was employed by ABARE to provide high-level independent advice. On 16 July 1997, four months before Kyoto, he resigned from his advisory position citing private-sector funding as posing 'major risks for the integrity and efficacy with which modelling work can be done'. He wrote that the problem was made especially severe when Government 'seeks to use results from a semi-secret proprietary model as a basis for justifying its policy position'. This constellation of circumstances, he wrote, 'is diametrically opposed to all that I have stood for during my 30 years as a policy modeller. To continue on the Steering Committee of MEGABARE ... would be hypocrisy of a high order'.[22]

Late in 1997 the Commonwealth Ombudsman launched an investigation of the funding arrangements of the steering committee. The ombudsman's report concluded that by limiting membership of the committee to organisations willing and able to pay $50,000, ABARE had failed to protect itself adequately from 'allegations of undue influence by vested interests'. With masterful understatement, it said that its practices 'could create a reasonable public perception that the research projects were weighted in favour of the interests of Australian industry'. It stressed that ABARE had misled readers of its reports by failing to acknowledge the financial contributions of industry, that the Government's climate change analysis was 'compromised', and that ABARE management had displayed 'poor judgement'.[23]

As the attacks on the credibility of the modelling reached a climax in the weeks before Kyoto, one of the fossil-fuel company members of the steering committee was heard musing over whether his organisation should 'ask for our money back'.[24]

DFAT GETS RELIGION

Throughout 1996 and especially 1997, in the lead-up to Kyoto, the Department of Foreign Affairs and Trade was dominated by a rigid view of climate change policy, in which the 'national interest' became indistinguishable from the commercial interests of the fossil-fuel industry. Officers working in the climate change area of DFAT at the time noted an atmosphere of 'religious fanaticism' in which no dissent or questioning was tolerated. The head of the climate change branch, Meg McDonald (who would later become a fossil-fuel company lobbyist), approached her task with unusual zeal. The branch experienced severe morale problems. Many foreign affairs officers join the service because they want to work at jobs that help make the world a better place. Those charged with arguing the Government's climate change position felt that they were, in the words of one, 'doing the Government's dirty work' and became ashamed rather than proud of their work. The staff turnover rate was said to be 'huge'.[25]

The Europeans were clearly identified by the Government and DFAT as the enemy. At a briefing in Canberra after Kyoto, an ABARE officer revealed the peculiar world view that underpinned the Government's approach. When asked to describe the 'Umbrella Group' (consisting chiefly of Australia, the United States, Japan, Canada and Russia), he said that it represented the 'free world' – as if nine years after the fall of the Berlin Wall, the Cold War were still being fought, except that Western Europe had become the enemy. The European Union argument for a uniform 15 per cent cut in emissions was seen by the Government as a severe threat to Australia's continued prosperity. In this febrile atmosphere, Australian environmentalists who criticised the Government's position were viewed as traitors. Astonishingly, diplomats abroad were instructed that they were not to communicate any criticism of the Australian position to the Government; ministers were only interested in good news. There was an in-house DFAT joke that said that all inbound cablegrams must begin with the words 'Australia's objectives were fully achieved'.

As a result, the Australian Government and its climate change advisers lost touch with reality. It was a disconnection that had been some time coming. Beginning with the merger of the departments of trade and foreign affairs to become DFAT in 1987, the goals of Australia's diplomacy had become increasingly skewed towards the promotion of trade interests at the expense of broader concerns, such as global environmental issues, human rights and peace. With the 'national interest' so redefined, the Government could repeatedly argue that any agreement to reduce our greenhouse emissions was 'against the national interest', as if Australians had no interest in contributing to the protection of the world from the effects of climate change. Within the bureaucracy the attraction of tangible and immediate benefits, such as mineral exports, increasingly overshadowed less tangible long-term objectives.

In the case of climate change, this fixation on 'export wins' took the form of seeking to protect one particular form of export. Australia's global economic prospects would be anchored in the export of energy and embodied energy to Asia. The forthcoming Kyoto conference was characterised as a threat to this grand vision.[26] Thus Meg McDonald expressed the view that the climate change negotiations were just international trade negotiations under a different name.[27]

The overwhelming importance attached to energy-intensive exports as the road to economic prosperity had some untoward consequences. First, it consigned to the rubbish heap visions of Australia's future being built on technological sophistication, so that talk of the 'clever country' became idle; businesses developing renewable and energy efficiency technologies searched in vain for a sympathetic ear in Canberra. Secondly, it ceded enormous political influence to energy-intensive export industries, with executives having free and frequent access to the relevant ministers, and the ministers making it clear to their bureaucrats that the industries' views were the Government's views, to the point where it was not regarded as improper that the fossil-fuel industries oversaw the formulation of greenhouse policy modelling.

In the mid-'90s, the system of government broke down in the development and prosecution of climate change policy. The various arms of the bureaucracy were captured by a narrow section of Australian industry – the greenhouse mafia – and hence failed to provide a balanced assessment of what was in Australia's interests, to the neglect of the broader economy and the country's interests beyond short-term mineral exports. Within the Government and the public service, the official position became unchallenged dogma and the provision of frank and fearless advice was impossible. Ultimately, the fanaticism with which the Federal Government pursued its position was damaging not only to attempts to arrest the global problem of climate change but also to the system of governance in Australia.

LAST MOVES

Meanwhile the fossil-fuel lobby was actively pushing the Government to harden its position even further. The staging of a conference on 19–21 August 1997 called 'Countdown to Kyoto' intensified the feverish atmosphere. The conference, held in Canberra, was convened jointly by the Australian Asia-Pacific Economic Co-operation (APEC) Study Centre, based at Monash University, and the Frontiers of Freedom Institute. The APEC Study Centre is chaired by Alan Oxley, a former Australian ambassador to the General Agreement on Tariffs and Trade who has been one of the foremost anti-greenhouse activists in Australia. He is a contributor to and co-host of Tech Central Station Asia-Pacific, perhaps the most effective denialist website and one, among others, funded by ExxonMobil.[28] Sponsors for the conference included Xstrata and ExxonMobil.[29]

The Frontiers of Freedom Institute is a far-right US think-tank funded by ExxonMobil and tobacco companies.[30] Even the conference's highly confidential media strategy, prepared by public relations firm Hannagan Bushnell and leaked to environment groups, noted that the 'backing of Frontiers of Freedom and known US right-wingers make obvious targets for green counter-moves'.[†] A fundraising letter from Frontiers of Freedom declared that the aim of the conference was to 'offer world leaders the tools to break with the Kyoto treaty'. Fossil-fuel and aluminium corporations were among the sponsors.

Countdown to Kyoto featured prominent US anti-greenhouse science activist Pat Michaels and right-wing US politicians Senator Chuck Hagel, Congressman John Dingell and Senator Malcolm Wallop. Wallop, as chair of the Frontiers of Freedom Institute, co-chaired the conference. He is strongly pro-guns, wants a total end to all social security and believes that the American people are 'more patriotic' and more inclined to 'do what is right' than any

---

† The letter from Hannagan Bushnell accompanying the leaked media strategy for the conference was addressed to 'Mr Ray Evans, WMC Limited'. The fee for five days' work was $10,000 plus expenses, but WMC must have been short of cash because the letter from Noel Bushnell went on to say 'we accept your budget constraints and would be pleased to act for you for the offered fee of $6000'.

other people in the world.³¹ The conference was one of the first indications of how the far right in the United States had taken up the anti-greenhouse cause.

Deputy prime minister Tim Fischer and environment minister Robert Hill were on the conference program, although the Howard Government appeared to keep some distance from the meeting, fearing being tarnished by its extremism. The rapporteur was John Stone, former treasury secretary, Queensland National Party senator and One Nation supporter. Although the conference may have succeeded in stiffening the Government's resolve, it was a public relations flop, with Greenpeace mounting disruptive actions.

In August 1997 no fewer than seven of the Federal Government's most senior ministers travelled to Japan to lobby against uniform emission reduction targets.³² The Government made increasingly exaggerated claims about the economic effects of such a target on Australia, including that wages would fall by 20 per cent by 2020, that petrol prices would double, and that 90,000 jobs would be lost. These claims, based on unpublished 'research' by DFAT, were so far-fetched that they could be taken seriously only in an atmosphere of hysteria.³³ The numbers were orders of magnitude higher than those estimated by ABARE, which had itself engaged in serious exaggeration of the economic costs of abatement.

In November 1997, documents were leaked suggesting that Australia was preparing to withdraw from the Kyoto negotiations.³⁴ This known willingness to withdraw may have strengthened the Government's position, as it signalled to the rest of the world a reluctance to compromise.

On the eve of the Kyoto conference, the Prime Minister made a statement to parliament announcing a major policy initiative. The policy document, *Safeguarding the future: Australia's response to climate change,* was an eleventh-hour attempt to boost Australia's negotiating credibility, as it was apparent to everyone at home and abroad that Australia was doing almost nothing to restrain its growth in emissions. The Prime Minister began by saying that: 'We have ... made it plain that we are not prepared to see Australian jobs sacrificed' and that 'Australia's campaign for equity and realism has won wider support'. He went on to announce 'the largest and most far-reaching package of measures to address climate change ever undertaken by any Government in Australia'. Consistent with the hyperbole of the times, he could have added that it was the best package 'since the dawn of time', for past practice had set an extremely low benchmark.

The Prime Minister claimed that his measures would reduce the growth of Australia's emissions from 28 per cent to 18 per cent over the period between 1990 and 2010, but would not risk '90,000 potential jobs'. Reflecting the Government's continued inability to grasp the seriousness of the issue and the need to initiate a process of widespread structural change in the economy, he declared, 'We are prepared to ask industry to do more than they may otherwise be prepared to do', as if this were a bold step forward. He announced that the package would be funded to the tune of $180 million over five years, 'a significant sum by any standards'.[35] The gloss was taken off the message when it was pointed out that this amounted to the cost of one bus ticket per Australian per year, an amount that seemed to many to be entirely incommensurate with the seriousness of the problem.

Just prior to Kyoto, an opinion poll conducted by *The Sydney Morning Herald*/ACNielsen-McNair showed that 90 per cent of Australians were either 'concerned' or 'very concerned' about global warming; 79 per cent felt that Australia should sign a treaty to cut emissions; and 68 per cent said economic pain should not stop such a treaty being signed.[36] Perhaps more disquieting for the Government, a survey of 2200 Australian company directors showed that nearly half favoured legally binding global reduction targets.[37]

6. DRAMA AT KYOTO

THE KYOTO NEGOTIATING STRATEGY

After the frenzy of the lead-up, the Australian Government's approach to the November 1997 Kyoto negotiations was surprisingly low-key. It could afford to take a back seat, for the conditions of Australia's participation in any final agreement had been communicated unambiguously to the rest of the world. Australia had made it very clear that it would sooner walk out than sign up to a deal it found unacceptable. This was a powerful bargaining chip for a country that had decided to discard its concern for diplomatic respect. In UN processes, consensus is a *sine qua non* for an agreement to emerge. Moreover, it would be almost impossible to enforce obligations to cut emissions if a country like Australia – rich and with the highest per capita emissions in the industrialised world – refused to co-operate. The prospects for bringing developing countries into the target-setting process at a later date would be undermined.

The Howard Government understood this. At the same time, it attempted to spread fear among developing countries even about an agreement that absolved them of the need to cut their emissions. On 28 November, as the negotiations began, ABARE issued a research report with an accompanying media release headed: *Developing countries the losers in emission abatement policies*.[1] It claimed that if developed countries were required to limit their emissions, developing countries would lose export markets, pay more for imports and see foreign investment decline: a triple whammy. At the same time the Government was arguing that any agreement that did not include developing countries would see them obtain an unfair competitive advantage. As so often, the Government wanted it both ways and was not embarrassed to claim that two wholly contradictory outcomes would occur.

Leading the delegation, Senator Hill arrived in Kyoto carrying a set of instructions. He was to hold out and refuse to sign anything that did not include Australia's two key demands: provision for a large increase in emissions, and the inclusion of emissions from land-clearing in the figures for the base year, 1990.

The Prime Minister had said in his parliamentary statement of 20 November that the 'far-reaching package of measures' could limit Australia's emissions to no more than 118 per cent of those in the base year, and that that was the best Australia could do. Yet it was made quite clear from the outset by the Kyoto conference chair, Raoul Estrada, that Australia would get nothing like the 'headline' increase of 18 per cent it sought, with a maximum increase for any country of 10 per cent at the very most.[2] This was where the second demand came in.[†]

By pure coincidence the rate of land-clearing in Australia had spiked in 1990; in the late 1980s it stood at around 500,000 hectares per year and from 1991 was about 325,000 hectares. But in 1990 it was a whopping 675,000 hectares, which added over 100 million tonnes of carbon dioxide to Australia's emissions (see Figure 4, p. 72).[3] This spike in the rate of land-clearing – due to peculiar factors in

---

[†] This has subsequently been confirmed. In evidence to the Senate Inquiry into Global Warming on 9 March 2000, the head of the AGO Ms Gwen Andrews said that the Government expected Australia's emissions to reach 118–120 per cent of 1990 levels in the commitment period 2008–12. When asked how Australia would meet its 108 per cent Kyoto target, Andrews said that the Government would have recourse to falling emissions from land-clearing (Australia, Senate, *Debates*, 2000, p. 7).

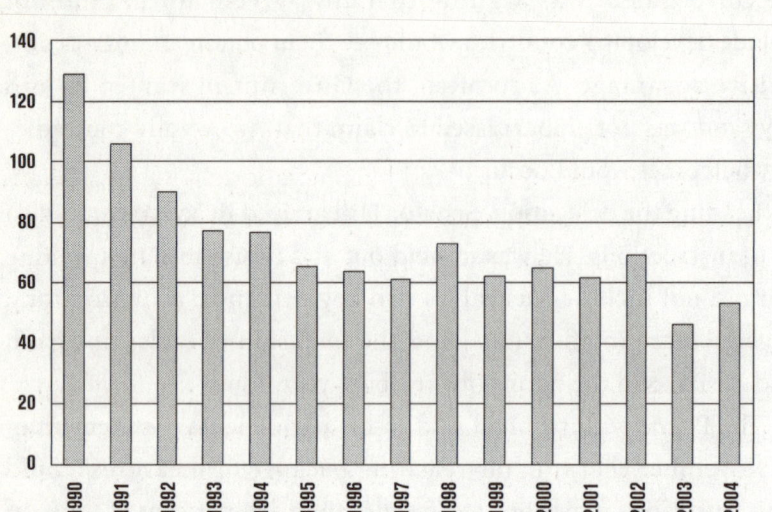

FIGURE 4: Greenhouse gas emissions from land-use change, 1990–2004 (Mt CO_2-e)

the economics of beef cattle – had a potentially profound implication for Australia's Kyoto target.

Australia is the only industrialised nation still to engage in widespread land-clearing (mostly in Queensland, although there is some in New South Wales), so its demand to be allowed to include emissions from land-clearing in its greenhouse accounting was immediately dubbed 'the Australia clause'.

Although he has subsequently denied it, Hill had in his possession modelling of the implications of inclusion of land-clearing for Australia.[†] As emissions from land-clearing had declined sharply since 1990, their inclusion in the base year would give us a cushion of 'free' emission reductions. Our fossil-fuel emissions would be able to increase to at least 120 per cent of 1990 levels by 2010 while

---

[†] Professor Graham Farquhar from the Australian National University, an expert in the terrestrial carbon cycle and a key person in compiling the land-use change and forestry portion of Australia's inventory, provided the modelling. Professor Farquhar was a very late inclusion in the delegation and was on hand to advise on details of land-clearing and forestry issues.

still coming in under an overall target of 105 to 110 per cent. The Australia clause represented a loophole in the Kyoto Protocol so big that a couple of bulldozers with a chain between them could be driven through it.

While Australia kept a low profile until the last hours of the Kyoto negotiations, political and bureaucratic activity in Canberra was intense. Australian negotiators provided daily updates of the positions being advocated by various parties, and the experts in Government departments used their spreadsheets to work out the implications for Australia. Canberra then indicated whether each proposal was acceptable or not. Within Government a long-running struggle between the environment minister on one side and the industry and foreign affairs ministers on the other was still being played out as the negotiations began. Over the two weeks of the conference, Cabinet met every other day, and more often as the finale approached. When the Prime Minister held informal drinks at the Lodge for Cabinet ministers, Kyoto was the hot topic. Bureaucrats were not invited, but Robert Hill managed to smuggle in Roger Beale, the secretary of his department, who posed as the barman! Carrying the drinks, Beale buttonholed ministers to put the environment department's view.[4]

In Kyoto, Hill had a strategy for winning public support at home for the Government's position. He held frequent private briefings for the Australian media contingent, from which foreign journalists and other Australians were excluded. This appeared to be an attempt to exploit the patriotism of media representatives and to persuade them, by taking them into its confidence, that the Government's position must be in the national interest. The Government also maintained that the national interest was being threatened by powerful opponents who, while appealing to environmental imperatives, were actually pursuing their own trade interests through subterfuge.

Throughout the negotiations the fossil-fuel lobby kept the pressure on the Australian delegation not to weaken its resolve. Those behind the Countdown to Kyoto conference sent the aggressive PR operative and former Alcoa employee John Hannagan to Kyoto, where he heavily lobbied the Australian media.[5] Hannagan issued a stream of media releases attacking the Europeans and green groups. In the hot-house atmosphere most of the Australian media representatives allowed themselves to be persuaded by the spin put out by the Government and the fossil-fuel lobby.

The negotiations dragged on. Many delegates and observers began to doubt that any agreement would be reached. It was only in the last hours of the conference, after the clock had been stopped at midnight on Wednesday 10 December, that delegates and others began to sense the possibility of an agreement. Pressure to reach a deal was intensified by the fact that delegates had to vacate the building so it could accommodate another conference (one sponsored, ironically enough, by the world nuclear industry). Exhaustion added to the pressure, but a deal acceptable to all of the parties was finalised in the early hours of Thursday 11 December. As Chairman Estrada for the last time went through the agreed text clause by clause, Senator Hill rose to point out the omission of the 'Australia clause' in Article 3.7 and to insist on its inclusion. The rest of the world had only a vague sense of the implications of the clause, but was unwilling to allow the protocol to founder on a concession to Australia. The clause was agreed to at 1.42 a.m. Writing for *The Australian*, Robert Garran and Stephen Lunn captured the drama of the moment: 'So after Senator Hill's interjection, Mr Estrada added a new sentence to the clause, tailor-made to give Australia the escape hatch it was seeking ... These were the words which saved the conference and allowed Australia to join the protocol'.[6]

The Australian media were caught up in the Government's euphoria and carried headlines such as 'Emission accomplished', 'Australia's greenhouse triumph' and 'Our 1.42 a.m. greenhouse coup', thereby endorsing the Government's view that it had protected the 'national interest'. However, public comment following Kyoto was highly critical of the Government, with many letter-writers and radio commentators expressing shame that Australia had dragged the negotiations backwards on this crucial issue. Cartoonists were struck by the irony of an environment minister celebrating success at watering down an international environment treaty. Labor's environment spokesman Duncan Kerr prophetically described the task given to Australia by the Kyoto Protocol as a 'three-inch putt'.

Australia won its extraordinary concessions by threatening to wreck the consensus. While all countries negotiated with the national interest in mind, none of the others defined their interests so narrowly and to the exclusion of the problem of climate change itself. The executive director of the convention secretariat, Michael Zammit Cutajar, had early in the conference referred to every country except Australia being committed to its success.[7] Raoul Estrada stated explicitly that Australia had been allowed to have its way only in the interests of obtaining unanimous agreement.[8]

The Australian negotiating strategy was no surprise; the Howard Government had been threatening to withdraw for some months before the conference. If any larger power, or a small number of countries, had behaved in the same way as Australia, agreement would never have been reached. Australia therefore took advantage of the more responsible approach adopted by other countries and exploited the fact that consensus on mandatory targets by all industrialised countries was essential to obtaining an agreement.

INTERNATIONAL REACTION TO THE AUSTRALIAN DEAL

Australia's negotiating tactics, and the 'victory' they delivered, generated worldwide resentment. The chief European negotiator, Ms Ritt Bjerregaard, said that the outcome for Australia was a mistake, that Australia had made a misleading case and 'got away with it', and that this would not be forgotten.[9] The European Union's spokesman on environmental policy, Peter Jorgensen, said that the Australian increase was 'wrong and immoral. It's a disgrace and it will have to change'.[10] Some conservative US and Canadian commentators immediately began to ask why their countries had not won such concessions.[11] Leading developing countries were reported to be preparing to use the Australian precedent as the basis for a refusal to cut their emissions.[12]

It might be argued that while Australia's tactics were mean and tricky, the strategy worked. Australia got what it wanted. But being seen to be a good global citizen was sacrificed in the process. The coin of good citizenship is valuable in part because it influences how other nations deal with us. Some insiders believe that Australia's stance before and during the Kyoto conference has resulted in serious diplomatic damage, damage that has been manifested in subsequent negotiations. The failure of Australia's nomination of Professor Ivan Shearer as a judge on the International Tribunal for the Law of the Sea in July 1997 and the failure of Australia's candidate for the position of commissioner on the Limits of the Continental Shelf have been linked to Australia's stance on climate change by well-placed observers.[13] Japanese negotiators still raise the topic of Kyoto during negotiations over access to southern bluefin tuna fisheries. And during negotiations in July 2000 over an Australian proposal to establish a whale sanctuary in the Pacific, Japanese negotiators referred to the Australia clause in order to undermine our environmental credentials.

In a book published in 1999, German academics Sebastian Oberthur and Hermann Ott analysed the positions of the parties in preparation for and at the Kyoto conference. They bracketed Australia with OPEC and Russia as the principal obstacles to progress in the early negotiations.[14] Oberthur and Ott drew the following conclusions: 'The Kyoto targets surely have two main winners: Russia and Australia ... The considerable increase in emissions allowed to Australia ... has set a bad precedent for future negotiations, especially with regard to developing countries'.[15]

Curiously, while Senator Hill seemed jubilant about the 'Australia clause' on land-clearing immediately after the event, in later comments on Kyoto he seemed reluctant to mention it, to the point of being misleading. In a letter published in *The Australian* on 31 December 1997 and in a speech on 30 January 1998 to the Committee for the Economic Development of Australia, Hill did not mention land-clearing *at all*. Yet when members of the Australian delegation flew to Dakar in May 1998 for Kyoto follow-up discussions, the number one objective was to 'protect Australian clause in Article 3.7' and to 'counter EU arguments that LUCF [land-use change and forestry] is especially prone to uncertainty'.[16] So began the process of protecting the gains of Kyoto in private while talking down the extent of the victory in public.

THE ELEMENTS OF THE KYOTO PROTOCOL

The Kyoto Protocol set emission reduction targets, referred to as 'assigned amounts', for the industrialised countries, known as 'Annex B countries'. Taking 1990 as the base year, countries undertook to reach their targets in the first commitment period, 2008 to 2012. The agreement at Kyoto was seen by all participants as no more than the first small step on the path to large-scale reductions in global emissions. This point, though simple and obvious, is important to note in light of subsequent distortions of the protocol by hostile sources in Australia and the United States who have claimed that it will have only a small impact.

The targets for selected industrialised countries are shown in the table overleaf.

TABLE 1: Kyoto Protocol targets for selected Annex B countries

Country	Target (%)
Iceland	110
Australia	108
Norway	101
New Zealand	100
Russian Federation	100
Ukraine	100
Canada	94
Japan	94
Poland	94
United States	93
European Union	92

Three points are worth noting. First, while on the face of it the reductions are small – and environmentalists were disappointed that they did not go further – the proper comparison is with emission levels under 'business-as-usual'. Thus, emissions in the European Union would rise by at least 20 per cent between 1990 and 2010 if no action were to be taken. As a bloc the European Union countries are required under the protocol to cut their emissions from an expected 120 per cent to 92 per cent, quite a tough target when action could begin only in 1998 at the earliest.

Secondly, the targets set for the United States, Japan and the European Union – which together account for 70 per cent of the industrialised countries' emissions – differ by only 2 per cent, a very small variation for countries with markedly different emissions profiles. Only three of the 39 industrialised countries were permitted to increase their emissions: Australia (108 per cent), Iceland (110 per cent) and Norway (101 per cent).

Thirdly, the countries of the European Union negotiated as a bloc with a view to assigning responsibilities among themselves afterwards. Each country in the European Union was assigned a target of 92 per cent of 1990 levels, but the protocol allows parties to reach an agreement to fulfil their commitments jointly, a provision that allows the European Union to establish a 'bubble' in which targets vary by agreement as long as the overall target of 92 per cent is met. Some European Union countries, such as Greece and Portugal, have been allocated targets higher than Australia's; others must then have targets below 92 per cent to compensate. This became the basis for the European emissions trading system that got under way in 2005 and which is revolutionising the task of carbon reduction. The agreement also allowed for the establishment of an international emissions trading system.

AUSTRALIA'S LOOPHOLE

The Australian Government claimed that the world community had accepted its argument for 'differentiated targets'. Certainly, Australia received a much more lenient target than other countries, all the more so with the inclusion of the Australia clause, but it was not because the rest of the world accepted the validity of the Australian arguments. Quite the reverse: the Australian victory was achieved not by force of argument but by threat of withdrawal. The rest of the world had a very different conception of fairness from Australia's. For instance, the Norwegian delegation argued that parties should take on burdens proportional to their per capita emissions and their levels of wealth. Both of these principles – polluter pays and ability to pay – would have seen Australia assigned more *stringent* targets than most other countries, rather than more lenient targets. The Australian proposal for differentiated targets was seen as self-serving and not based on any recognised principles of equity.

It is only because of Robert Hill's insistence that the Australia clause be inserted into the Kyoto Protocol, and the fact that land-clearing has been declining from its historic high in 1990, that the Howard Government has been able to claim, right up to the present day, that Australia is 'on track' to meet its Kyoto target. This claim, repeated over and over, is designed to give the Australian people the impression that the Howard Government takes climate change seriously and that its policies are working to reduce emissions. It will now be clear that the claim is cynically deceptive.

Even so, it is still played as the trump card by the Government at every opportunity, as if it proves its commitment to cutting emissions. For example, the Australian Greenhouse Office inventory released in May 2006 seems to show that Australia's greenhouse gas emissions have increased by only 2.3 per cent since the Kyoto base year of 1990. If Australia had ratified the protocol, it would be required to limit the growth of emissions between 1990 and the 2008 to 2012 period to 8 per cent. The environment minister at the time, Ian Campbell, claimed in a press release that these figures vindicated the Government's policies and pointed to Australia's 'leading role' in various international processes. With astonishing brio, the minister declared that 'while the figures represent good news, we can't afford to be complacent'.

In fact, estimated emissions from land-clearing fell from a massive 129 million tonnes in 1990 to almost 53 million tonnes in 2004, a decline of nearly 60 per cent. This provided a cushion of 76 million tonnes of 'free' emission reductions exclusively available to Australia.

According to the Government's own figures, excluding land-use change and forestry our total emissions have grown by 25 per cent since 1990 (see Figure 5, p. 81), driven largely by the rapid increase in emissions from energy use (up by 35 per cent over the period).[17] By 2010, the expected increase in emissions excluding land-use

FIGURE 5: Changes in Australia's total greenhouse gas emissions with and without land-use change and forestry, 1990–2004 (Mt CO_2-e)

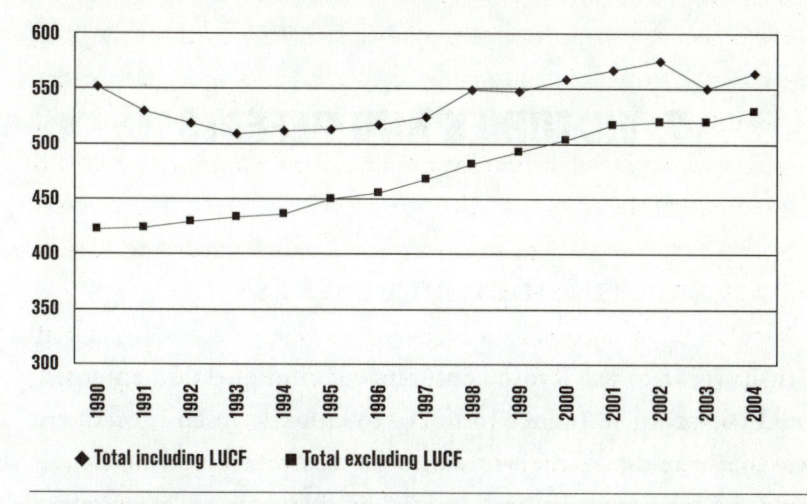

change will be over 30 per cent. This is the proper comparison with the targets accepted by other countries under the protocol.

The Government's claim to be 'on track' is a lie designed to give the impression that it takes climate change seriously and has policies in place that are cutting emissions to meet the Kyoto target. The claim is expressly designed to mislead the public. The deception was maintained through to the beginning of 2007, when new Australian figures showed that emissions from burning fossil fuels have been growing so rapidly that Australia is now expected to overshoot the 108 per cent target. The Government immediately began to change its story.

7. VICTORIES AND DEFEATS

THE JOLT TO BUSINESS

Within weeks of the Kyoto conference a vital question emerged: would the protocol induce industry to embark on an investment wave that would take the world into the next energy revolution, or would the agreement be undermined by filibustering, exploitation of loopholes and refusal to comply? The response of industry was pivotal to the success of the agreement.

The early signs were positive. Business opposition to emission reductions began to splinter. A growing number of oil company executives shifted away from their hard-line oppositionist stance and accepted the science on global warming. In the months after Kyoto, senior executives from British Petroleum (BP), Royal Dutch Shell, Texaco and Sun Oil made public comments indicating that they took climate change seriously and that oil companies would need to make substantial alterations. According to *The Washington Post*: 'One oil industry official recently warned his colleagues not to fall into the trap faced by the tobacco industry, which for years denied that cigarettes were addictive'.[1] The CEO of Sun Oil, the major refiner on the US East Coast, wrote in a letter to President Clinton that the scientific evidence was strong enough to justify 'prudent mitigation measures now'. He added a postscript: 'Hang tough, Mr. President – I believe the American people will be with you'.[2]

Early in 1998, a Texaco spokesman told financial leaders in Davos, Switzerland, that 'the debate isn't about the science anymore. It's about what companies are doing, and what they are doing is to look at the next generation of technologies'.[3] Others spoke of a generational shift in the values of their own workforces. A UK oil executive said that some of his company's young geologists were members of Greenpeace.

In April 1998, Shell Oil, the US unit of Royal Dutch Shell, withdrew from the Global Climate Coalition – the powerful fossil-fuel lobby group that had almost succeeded in derailing the Kyoto conference – citing irreconcilable differences over ratification of the protocol.[4] BP also defected from the coalition, thereby simultaneously earning the hostility of more conservative oil corporations and winning the plaudits of international environmentalists and enlightened policy makers around the world. John Browne, the company's CEO, said: 'We may have left the church in terms of climate change. But it is almost impossible to express the depth of support from within the company for the position we've taken'.[5] However, Exxon, one of the stalwarts of the group, while under pressure from its stockholders to take global warming more seriously, resisted change, and has remained recalcitrant ever since.[6] Mobil (which merged with Exxon in 1999) was another powerful oil company that decided to keep its head in the sand.

A sharp division emerged between oil companies in Europe and the United States. The willingness to take collective action to tackle environmental dangers is greater in Europe, and oil companies with headquarters there were much quicker to acknowledge that greenhouse gas emissions must be reduced. A month before withdrawing from the industry grouping, Shell senior executive Mark Moody-Stuart said: 'We have been repeatedly attacked in Europe for Shell Oil's membership of the Global Climate Coalition'.[7] Ford and Daimler-Benz left the Global Climate Coalition in early 2000, soon followed by Texaco, the first major US oil company to jump ship. Texaco declared that the company 'shares society's concern over the issue of climate change'. A growing list of US corporations, including DuPont, Alcan, Suncor Energy, IBM, Polaroid and Johnson & Johnson, made commitments to substantially reduce their greenhouse gas emissions. Some of these companies joined the moderate Partnership for Climate Action, which commits large energy and manufacturing corporations to reducing emissions below levels required by the Kyoto Protocol.[8]

The automobile industry was a crucial one. According to one view, there was a 'quiet revolution' under way in the industry, spurred initially by legislation in the 1990s requiring improved fuel efficiency and given a major boost by the protocol. But when the state of California enacted a landmark bill in 1990 imposing strict new limits on vehicle emissions, Detroit fought these restrictions every inch of the way. As one commentator observed: 'The auto companies were hiring lawyers to fight higher mileage standards while Japanese firms like Toyota were hiring engineers to design more efficient and environmentally-friendly cars'.[9]

Some in the US auto industry, however, began to accept that change was inevitable and turned their attention to designing more fuel-efficient vehicles. Ford is an interesting case. Under the chairmanship of William Ford – the great-grandson of Henry Ford – the corporation declared itself to be preparing for a world beyond the internal combustion engine. In 2000, it announced plans to improve the average fuel efficiency of its sports utility vehicles by 25 per cent over the next five years.[10] In that year Ford had to weather enormous flak from conservatives on Capitol Hill for its stance on greenhouse, but held its ground.

We should, of course, treat stories of revolution in the auto industry with caution, as it is frequently observed that Americans are addicted to gasoline. But vehicle technology is undergoing a transformation. Within the next decade or so, low-emissions technologies such as fuel cells will become commercially viable. Already hybrid vehicles such as Toyota's Prius model, with half the fuel consumption of other new vehicles, are being produced in large numbers at close to competitive prices. Honda, Ford, General Motors and Chrysler are all investing heavily in fuel cell, electric and hybrid vehicles. Not long after the Kyoto conference, one senior auto company executive predicted that by 2015 barely half the cars produced would have internal combustion engines (a prediction that now seems unattainable).[11] The president of General Motors was widely quoted as saying that the end of the internal combustion engine was in sight.[12]

To counter the enormous public opinion and lobbying effort still being financed by the fossil-fuel lobby and various right-wing organisations (including the Moonies, who owned *The Washington Times* newspaper, which took a consistent anti-greenhouse stance), the Pew Charitable Trust, a large US philanthropic organisation, set up a new policy centre aimed at disseminating more balanced views on climate change. By the end of 1998 it had support from Boeing, Lockheed-Martin, Toyota, Whirlpool, 3M, BP, Sun Oil, American Electric Power and Intercontinental Energy.[†] As well as accepting the science and likely impacts of global warming, these companies agreed that the Kyoto Protocol 'represents a first step in the international process, but more must be done both to implement the market-based mechanisms that were adopted in principle in Kyoto and to more fully involve the rest of the world in the solution'.[13]

Encouraging as these defections were, the corporate interests of some of the oil majors in particular provided a powerful counter to good intentions. The companies must be judged by their actions. While there is no doubt that oil companies such as Shell and BP are

---

† By 2006 its 42 members accounted for $2.4 trillion in market capitalisation and over 3.3 million employees. In addition to those above, they now included ABB, Alcoa, Bank of America, Duke Energy, Dupont, GE, IBM and Rio Tinto.

investing large sums in renewable energy, and auto makers are investing large sums in low-emissions vehicles, they also continue to invest much more in traditional activities. Shell, for instance, is still exploring for oil and gas in some of the world's most environmentally sensitive areas, and Ford continues to supply the burgeoning market for gas-guzzling sports utility vehicles.

There were other attempts to white-ant the Kyoto agreement. One of the more duplicitous of these was the so-called Leipzig Declaration of 1995, updated in 1997, in which almost 100 scientists from leading universities said that they could not subscribe to the view that climate change represents a serious threat. A Danish investigative television program showed the declaration to be largely fraudulent, with most of the signatories either not climate scientists, people with no scientific standing, scientists with standing who said they did not sign it or, in at least one case, a Florida television weatherman.

The Kyoto agreement gave a major boost to research and investment in various forms of renewable energy. Clean energy technologies – wind, solar, biofuels, geothermal and hydropower – have become mainstream in many parts of the world. For example, global investment in wind energy reached nearly US$12 billion in 2005, a 47 per cent increase from a year earlier; it is expected to reach US$49 billion by 2015.[14] Similarly, investment in solar photovoltaics had reached US$11 billion by 2005 (up 55 per cent from the previous year) and is projected to reach US$51 billion by 2015.

Even at the Fourth Conference of the Parties, held in Buenos Aires in November 1998, it was apparent that the world had changed considerably in the year since Kyoto. At previous meetings, the fossil-fuel lobby had been easily the dominant business group. The sustainable energy industries were present but acted like bit players, while the huge insurance industry issued ominous warnings that went largely unheeded. At Buenos Aires it was apparent that some of the biggest industries were taking a keen interest in the issue and revising their business strategies so as not to be locked out of the post-Kyoto world.

To the surprise of almost everyone, the world's business leaders meeting in Davos in early 2000 declared climate change to be the greatest threat facing the world. In November of that same year, a poll of Fortune 500 executives reported that 34 per cent supported ratification of the Kyoto Protocol, with only 26 per cent opposed and the rest undecided.[15] At the same time, *The Financial Times* reported that investor interest in renewable energy companies had 'swung dramatically' over the previous few years, with most interest in wind power systems.[16] It was no longer accurate to think of renewable energy producers as small players suited only for idealistic greenies.[†]

THE HAGUE FIASCO

While the views of business and industry were rapidly altering, the three years between the Kyoto conference and the crucial Sixth Conference of the Parties, at The Hague in November 2000, saw little progress on finalising the details at a government level. The more reluctant signatories banded together to form the Umbrella Group – the United States, Canada, Japan, Russia, Australia and a few smaller countries – which embarked on a search for loopholes in the protocol. (Australia, of course, already had exclusive access to a massive loophole in the form of the clause on land-clearing.) The Hague conference was widely seen as the make-or-break meeting at which key decisions would need to be made if the parties were to have time to implement policies to achieve their targets by 2012.

---

† In 2007 the four biggest companies in the solar power field are large enterprises – Siemens in Germany, Kyocera and Sharp in Japan, and British Petroleum in the United Kingdom.

However, in 2000 a new issue emerged, one that had the potential to open up the mother of all loopholes.

A part of the protocol that had attracted little public attention dealt with net emissions from 'additional activities' covering a range of agricultural practices. The IPCC calculated that if these potential exemptions were exploited to the maximum, it would absolve industrialised countries of *any* need to reduce the growth of emissions from fossil fuels.[17] The environmental integrity of the protocol would be destroyed.

Even so, no-one was prepared for the breathtaking ambit claim lodged by the United States in early October 2000, in which it called for so wide a definition of additional activities that it would allow the United States to meet its Kyoto commitment without reducing domestic emissions. So contrary was it to the spirit of the Kyoto Protocol that it could only be interpreted as an aggressive negotiating tactic, one that set the boundaries between the European Union and the United States so far apart that any meeting in the middle would be to the great advantage of the latter (although 'advantage' is used advisedly, since it can hardly be said that it is in the long-term interests of US citizens to see almost nothing done to reduce global emissions).

In the end it was the ambit claim put by the United States that caused The Hague negotiations to implode. The delegates went home having achieved nothing. Searching for a scapegoat, the head of the US delegation characterised the breakdown as 'a crisis of European governance', but the common European view was summed up by the Portuguese environment minister: 'I feel it is better not to have an agreement than to have a bad agreement'.[18] After all, if an already weak agreement were now to include loopholes that meant major polluters needed to do very little to cut their emissions, the public would be gulled into believing that something was being done about climate change when in fact very little was. A watered-down Kyoto Protocol might do more harm than good.

AFTER THE HAGUE

Although The Hague conference broke up in disarray, the possibility of going ahead without the United States soon began to be canvassed. If nations accounting for 55 per cent of the 1990 carbon dioxide emissions of industrialised countries were to combine, that would permit the protocol to enter into force and become legally binding on all ratifying countries. If this happened, US corporations could be shut out of the Kyoto trading system and the mechanisms encouraging investments in developing countries. In addition, if the United States continued to refuse, other countries would be compelled to take measures to protect their trade interests.

The Hague Conference collapsed in the last days of the Clinton Administration, but it did so under the shadow of George W. Bush's imminent presidency. Although promising to cap greenhouse gas emissions from power plants if elected, President Bush immediately took a hard-line stance on climate change, repudiating his election promise and declaring in March 2001 that the Kyoto Protocol was 'fatally flawed' and that the United States would not ratify it. International condemnation of Bush's arrogant unilateralism was immediate, with the European Parliament declaring itself 'severely disappointed' by the 'unilateral and non-co-operative' position. One member of the UK Parliament dubbed Bush the 'toxic Texan'.[19] Furious opposition came not just from European political leaders and commentators but also from many corporate leaders. In the words of *Business Week*, 'his Administration was caught off guard by the negative reaction he has been receiving from corporate groups – usually a Republican's best friends'.[20] A professional poll conducted by ABCNEWS after Bush's statement on Kyoto showed that 61 per cent of Americans, including 52 per cent of Republicans, answered 'yes' to the question: 'Should the United States join the Kyoto treaty?' Only 26 per cent said 'no'.[21]

Bush surrounded himself with greenhouse sceptics. Vice-president Dick Cheney was already an implacable opponent of Kyoto, and the President appointed climate change conservatives to key positions – notably Gale Norton, described by some as an 'undercover operative' for oil and mining interests, as interior secretary. However, Colin Powell, Bush's secretary of state, called for a delay in resumption of talks after The Hague so that the new Administration would have time to make a 'constructive contribution'. An interesting appointment was that of Paul O'Neill as treasury secretary. As the CEO of Alcoa, O'Neill had for some years been arguing that governments must act to reduce emissions and supported increases in petrol prices through the Clinton–Gore carbon tax of 1992. O'Neill strongly supported a principled approach, saying:

> If people believe this is a serious issue, they ought to put their nationalism behind them and think about how does civilisation exist with this problem we're creating. Then we'll sort out who pays. Seems fairly obvious to me. People with money are going to have to pay ... If we really care about this as a global problem, we and other developed nations are going to have to pay for it.[22]

O'Neill took a proposal to increase fossil-fuel taxes to the first meeting of the Bush Cabinet, but it was given short shrift. In March 2001 Bush publicly repudiated the Kyoto Protocol. O'Neill was out on a limb and was soon forced to resign.

Under the influence of denialists, the hard-liners in the Administration never accepted the science of climate change, with the President even referring to the reports of the IPCC as 'foreign science' (although at least a third of the scientists involved were American). In May 2001, the White House requested that the US National Academy of Sciences prepare a review of the science. In June, 11 atmospheric researchers who were members of the academy, including the sceptic Richard Lindzen, affirmed the mainstream scientific view.[23] They concluded that the Third Assessment Report was an

> admirable summary of research activities in climate science ... The IPCC's conclusion that most of the observed warming of the last 50 years is likely to have been due to the increase in greenhouse gas concentrations accurately reflects the current thinking of the scientific community on this issue.

When the Sixth Conference of the Parties resumed in Bonn in July 2001, US repudiation seemed to have galvanised the other parties into avoiding failure, and they resolved most of the outstanding issues. The rest of the world seemed determined to prevent the United States from undoing a decade of hard-won progress. Delegates cheered and hugged each other as the gavel came down. UK environment minister Michael Meacher described it as 'a brilliant day for the environment' and a spokesperson for G77 said it was 'a triumph of multilateralism over unilateralism'. The head of the US delegation, Paula Dobriansky, was booed by delegates. At home, the Bush Administration faced a storm of criticism, with the powerful Senate Foreign Relations Committee voting 19 to zero on a motion declaring that the United States should not 'abandon its shared responsibility to find a solution to the global climate change dilemma' and calling on the Government to take actions to ensure 'significant and meaningful' reductions in US emissions.

THE SHRINKING COMMONWEALTH

International climate change disputes came to Australian shores in March 2002, when the Howard Government hosted the Commonwealth Heads of Government Meeting. For many Commonwealth countries, climate change is the most serious threat to their long-term prosperity. Indeed, the very survival of some nations is at stake, with scientific forecasts indicating that some low-lying areas, including whole islands, will be inundated by rising sea levels. In the Pacific, Tuvalu, Kiribati, Nauru and Tonga (all Commonwealth countries) are the most vulnerable.

The Howard Government dismissed the scientific consensus and the concerns of Pacific leaders about sea-level rise, and referred to the work of the National Tidal Facility, an Australian aid-funded project that had been measuring sea levels in the Pacific since 1994. In fact no conclusions about sea-level rise could be based on this data as it involved only a handful of measurements. Prime Minister Talake of Tuvalu told delegates to the meeting that the evidence of his own eyes debunked Australia's arguments – islands on which he had played as a child had either sunk or were sinking. Extreme tides in February 2001 were higher than he had ever experienced in his 60 years.[†]

The 2001 IPCC report had projected declines in crop yields threatening food security and thus the basis of life support in many countries. Due to warming and changing rainfall patterns, crop yields are expected to decline in most Commonwealth developing countries, including throughout Africa, where 350 million citizens of the Commonwealth will be affected. Drastic declines in yields (of 20 to 30 per cent) are also expected for India and Pakistan, affecting more than 1 billion Commonwealth citizens.

Following the 2001 'boat people' election, it was pointed out, not

---

[†] The then head of the National Tidal Facility, Dr Tad Murty, was one of the signatories of a 2006 open letter to the Canadian Prime Minister that challenged the science of climate change. His name was among those of most of the well-known denialists, including Fred Singer, Patrick Michaels, Richard Lindzen and Sallie Baliunas, and Australian sceptics William Kininmonth, Ian Plimer and Bob Carter. The letter is carried on most denialist and sceptic websites, including that of the Lavoisier Group: <http://www.lavoisier.com.au/papers/articles/canadianPMletter06.html>.

least by former US president Bill Clinton while on a visit to Sydney in that year, that climate change is expected to force many people in developing countries to move out of their homelands. In contrast to a trickle of asylum seekers, it is possible that in future a flood of environmental refugees will demand entry to Australia. Indeed, late in 2000 a senior Tuvalu official, Mr Paani Laupepa, told the BBC that Tuvalu had approached New Zealand and Australia to accept people displaced by rising seas. 'While New Zealand responded positively in the true Pacific way of helping one's neighbours, Australia on the other hand has slammed the door in our face.'[24] When asked about Tuvalu's concerns late in 2001, immigration minister Phillip Ruddock made the peculiar claim that accepting environmental refugees from Tuvalu would be akin to a return to the White Australia Policy. Having refused point-blank to countenance accepting environmental refugees from Tuvalu, a few weeks later the Australian Government approached Tuvalu to accept Middle Eastern asylum seekers turned away from Australia.[25] A spokesman for the Tuvalu Government said it would consider any request but that his country lacked the space, adding: 'We ask them for space and now they're sending us their own people'.‡

The Howard Government had been working actively for some years to stifle protests from small island states in the Pacific. At the 1997 South Pacific Forum, Australia had threatened to reconsider its aid budget unless a resolution on the subject of environmental refugees was watered down. In response, Howard dismissed the concerns of Pacific island states as 'exaggerated' and 'apocalyptic', and

---

‡ In early 2006 the Labor Party announced a policy of accepting refugees from Pacific Islands escaping rising sea levels, nominating Tuvalu, Kiribati and PNG's Carteret Islands as most at risk from salt-water intrusion and storm surges. *The Australian* sought a response from anti-Kyoto economist Warwick McKibbin, who opposed the idea, thoughtfully declaring: 'If it stops snowing in New Zealand, does Australia then take all the New Zealanders?' (Amanda Hodge, '"No room" for ALP's climate refugees', *The Australian*, 6 January 2006).

said that 'the jury is still out' on climate change.[26] Perceptions of the crudeness of the Australian position were reinforced when comments by the Government's chief economic adviser on climate change were circulated at the forum. Dr Brian Fisher, executive director of ABARE, had told a conference in London that it might be more efficient to *evacuate* small island states subject to inundation rather than require industrialised countries like Australia to reduce their emissions.[27] In 2001 at the Pacific Island Forum in Samoa, a group of 15 nations tried to include reference to 'environmental refugees' and compensation for people affected by climate change in a statement to be sent to the World Summit on Sustainable Development in Johannesburg in 2002. Australia insisted that the reference be removed and, after a tense standoff, the group of 15, including New Zealand, backed down.

RUSSIAN RATIFICATION

By 2004 the Kyoto Protocol had been ratified by all of Western Europe, Canada and Japan. Everything now hinged on Russia's decision, which if favourable would get the numbers up to the requisite 55 per cent of industrialised country emissions. Russia had gained a huge concession at the Kyoto conference. Since 1990, the Kyoto base year (which was one year after the dissolution of the Soviet Union), much of the old inefficient Soviet industry had collapsed, causing a sharp decline in Russia's greenhouse gas emissions. In fact, emissions were down by around 25 per cent by 1997. The target set by the protocol for Russia's emissions in the 2008 to 2012 commitment period was a zero per cent increase on 1990 levels. Even with good economic growth, Russia's emissions would be unlikely to reach this level, leaving Russia with surplus emission credits that it could sell to other Kyoto countries struggling to meet their targets. This gap in Russian emissions was dubbed 'hot air' and represented a financial windfall to Russia. The Ukraine was in a similar position. Those concerned with ensuring the environmental effectiveness of the protocol were worried that Russian hot air would provide the United States and Japan in particular with a form of 'offshore compliance', such that they could buy their way out of much of the domestic emissions reductions that their own Kyoto targets seemed to demand. As one commentator observed, 'American Cadillacs will be fuelled by Russian depression'.[28]

When the United States announced it would not ratify, the largest market for Russian hot air disappeared. In the meantime, battle lines were drawn in the Kremlin between hard-liners and sceptics, who repudiated Kyoto, and modernisers, who wanted greater integration into the world economy, warmer relations with the European Union and more influence in international affairs. The environmental implications of climate change never seemed to figure prominently in Russian calculations. Throughout 2003 and 2004 mixed signals emerged from Russia, but behind the scenes the crucial issue was gas prices. Russian gas prices were around one-fifth of those in the European Union, and Europe made equalisation of prices a condition for its support of Russia's application to join the World Trade Organization. Following a complex deal in which Europe settled for a doubling of Russian gas prices, President Putin announced in May 2004 that his country would 'rapidly move towards ratification'. Putin's announcement injected a life-saving shot of adrenaline into the international process, although most observers decided to wait for the vote in the unpredictable Duma before they celebrated. But on 22 October 2004 the Duma voted in favour and on 16 February 2005 the Kyoto Protocol, seven years after it was agreed, finally entered into force. The world would never be the same.

8. AUSTRALIA: AFTER KYOTO

A NEW BUREAUCRACY

Let me now turn to the parallel political developments in Australia. At the first Cabinet meeting after his triumph in Kyoto in 1997, environment minister Robert Hill received a standing ovation from his colleagues. The senior public servants who had worked so hard to prosecute the Australian case were rewarded for their loyalty. The leader of the Australian delegation, Meg McDonald, was promoted to a senior post at the Australian Embassy in Washington. In recognition of the role its modelling played – despite being discredited for taking money from fossil-fuel companies – ABARE received a special award for public service from the Prime Minister. The fossil-fuel lobby thought that Australia had done particularly well, with the then executive director of the Australian Industry Greenhouse Network, Tony Beck, writing to the Prime Minister on 22 December 1997 in the following terms:

> I am writing to convey to you directly our congratulations on the excellent outcome in Kyoto. Your personal and long-standing commitment to an equitable outcome for Australia, together with the highly effective 'whole of government' approach to the negotiations, has paid off handsomely against considerable odds.[1]

After the Kyoto celebrations, the commitments required some action. The establishment of the Australian Greenhouse Office (AGO) had been announced by Howard in his statement of 20 November 1997. Initial comments from the Government gave the impression that the office would bring together all greenhouse policy activities in the Commonwealth Government under the environment minister, and greenhouse matters would be co-ordinated from that portfolio.

However, in an early sign that the man trusted to negotiate the deal would not be trusted to follow it through, Hill was over-ruled. The months between December 1997 and April 1998 when the Australian Greenhouse Office commenced operations were marked by a bureaucratic turf battle prompted by those who saw dangers to their interests in the environment portfolio controlling a powerful policy agency. In a rebuff to Senator Hill, policy responsibility for the AGO was entrusted to a ministerial council of three ministers, those for environment, primary industries and energy, and industry, science and technology.

By this arrangement, the influence of the environment department over greenhouse policy was immediately and deliberately weakened. While all greenhouse activities of the department were moved into the new AGO, only some in the industry and resources departments moved and others remained where they were. The ultra-dry and sceptical Senator Nick Minchin, who was appointed industry minister in October 1998, retained a direct responsibility for the Greenhouse Challenge Program. This was seen as an excellent opportunity by the greenhouse mafia. The head of the Pulp and Paper Manufacturers Federation, Bridson Cribb, emailed the inner circle of mafiosi in October outlining concerns about the introduction of emissions trading and the mooted 2 per cent renewables target, and suggesting that having Minchin on side would enable them to use his department 'to (at the least) counterbalance the role of

Roger Beale, EA [Environment Australia] and the AGO'.[2] The Australian Aluminium Council had already set aside $75,000 in 1997 for a 'rapid response' strategy to widen loopholes in the Kyoto Protocol and influence domestic policy.

While the influence of the Australian Greenhouse Office on the development of policy was for a time strong, it became the subject of relentless attacks from the fossil-fuel lobby and was undermined by other ministers, the more so as it became clearer that the Government had no interest in programs that would constrain the polluting industries. Even before that happened, there was internal confusion. Any significant policy move by the AGO had to be approved by the ministerial council. The other ministers had their own resources to develop greenhouse policies. Thus a Cabinet submission prepared by the AGO was the only way the environment minister could secure Government endorsement for new greenhouse measures, yet such a submission could be vetoed by his fellow ministerial council members. If it did get through, they would have warning beforehand and could block it in Cabinet itself, an event that occurred regularly.

Climate change policy became the focus of ministerial power struggles, especially between Hill and Minchin, factional rivals from South Australia. Hill had been intensely involved in the issue since he took over the portfolio in 1996, and understood the international commitments Australia had made. In a sense, he had something at stake personally, for it was he who had made promises across the negotiating table at Kyoto. Minchin had the advantage of being close to Prime Minister Howard but was a latecomer to the issue and, like all latecomers, had to be educated away from the narrow perspective of the fossil-fuel industries. The two ministers fought it out through policy proposals, notably the Mandatory Renewable Energy Target and the proposed insertion of a 'greenhouse trigger' in the new *Environment Protection and Biodiversity Conservation Act*.

The Government appointed Gwen Andrews to head the Australian Greenhouse Office. With a bureaucratic background in the Department of Finance and an unassuming manner, Andrews was probably useful early on in allaying concern in industry at the creation of the new office. However, as the AGO suffered one Cabinet defeat after another, the hopes of the staff to be part of Australia's response to the world's biggest environmental threat were deflated and morale fell. Andrews resigned in 2002 and later said that over her four years in the job she was not once asked to brief the Prime Minister on the issue.[3] The fact that Howard was at no time interested in hearing from his most senior greenhouse adviser spoke volumes for how the debate was unfolding in Canberra.

WINDING BACK KYOTO

In April 1998 Robert Hill travelled to UN headquarters in New York to formally sign the Kyoto Protocol (the stage before ratification), yet even at that point plans were afoot to see that it would never be ratified. The agreement had been a major defeat for the fossil-fuel lobby in Australia and it immediately began a process of undermining the commitment of the Government to the deal that it had struck. That the Government never really accepted that climate change was a problem, or that it would require far-reaching changes in economic policy, became evident in August 1998 when it released its long-awaited proposal to introduce a goods and services tax. The policy package proposed to cut the price of diesel fuel by 25 cents per litre for heavy vehicles. The price of petrol would fall by 9 per cent for all business vehicles, and the price of cars and light commercial vehicles would be cut by 16 per cent. Prices of public transport tickets would rise by 10 per cent and the incentive to convert cars from petrol to gas would be diminished. All of these measures would stimulate the use of fossil fuels, yet the tax package made no mention at all of the environment. When asked about the environmental implications of the new tax system, the Treasurer, Peter Costello, was genuinely mystified.

The Government began to advance arguments against the Kyoto Protocol. Within months the line began to be heard, imported from the United States, that developing countries had been 'exempted' from the protocol. This was 'unfair', it was claimed, and it undermined the effectiveness of the agreement because developing countries would soon be responsible for more emissions than rich countries. This was the first line of attack from the fossil-fuel lobby, and it was one that resonated strongly with the Government.

As explained in Chapter 2, it was an incorrect and insulting claim. Climate change is caused by increased *concentrations* of greenhouse gases in the atmosphere, and around 75 per cent of the current increased concentration is due to the activities of developed countries in the process of growing rich. It will be perhaps 50 years before developing countries match the pollution created by the rich countries over the last two centuries. In per capita terms, developing countries typically have one-tenth to one-twentieth of the emissions of the United States and Australia. Australia's 20 million people produce more greenhouse pollution than Indonesia's 200 million.

Moreover, every international agreement on climate change – the 1992 Framework Convention, the 1995 Berlin Mandate and the 1997 Kyoto Protocol – explicitly recognises that developing countries will be required to cut their emissions, but only after rich countries have led the way. This view was based on the widely accepted principles of polluter pays and ability to pay, according to which wealthy countries like Australia with high emissions should do much more than poor countries with low emissions. Yet the Howard Government began to vilify poor countries for failing to pull their weight. In 2001 Foreign Minister Downer expressed the contempt with which the Government viewed the concerns of developing countries when he declared: 'It is no solution at all ... if China and India and Brazil can go ahead and pollute the environment to their heart's content because we're all feeling a bit sorry for them'.[4]

The next argument the Government used was that the agreement was not in Australia's 'national interest', a slippery concept that had come to mean whatever the Government wanted it to mean. By now it had become clear that, more than any previous one, the Howard Government believed that the national interest was the same as our short-term economic interests, as if we had no national interest in being part of international attempts to tackle the most severe environmental threat facing the globe. Even so, the economic effects of Kyoto were grossly exaggerated by the Government.

Having written to the Prime Minister congratulating him on the 'excellent outcome in Kyoto', it was not long before the fossil-fuel lobby set about eroding the agreement. Within months of the conference, the Australian Industry Greenhouse Network began examining the implications of the Australian 108 per cent target and the likely role of emissions trading. By August 1998 one of the key members of the fossil-fuel lobby, John Eyles, was writing to other members arguing that 'we are in danger of losing our way at the strategic level of greenhouse policy'. He worried that the Government was getting mixed messages from industry – that industry supported the Kyoto outcome and was 'prepared to acquiesce' in its ratification.[5] Industry should state its position unambiguously: the Government should not ratify unless major developing countries were given commitments and that in the meantime greenhouse policy should confine itself to 'no-regrets' measures. Eyles recognised that asking for a retreat from the Prime Minister's statement would have political risks for the Government, but urged a strategy of calling on it to refuse to ratify until everyone else did.†

---

† By April 1999, Eyles was writing confidentially to the inner circle that 'we are being driven to adopt reactive and defensive positions', and that they needed to regain the initiative (John Eyles (AIGN), faxed memo, 8 April 1999).

THE REPUDIATION

The Government seemed unable to grasp how the world was changing after Kyoto, and entered a long period of policy paralysis. On 26 September 1998, newspapers reported leaked documents indicating that the Howard Government had made a secret Cabinet decision not to ratify the Kyoto Protocol unless the US Government did so first. The story ran on the front page of *The Canberra Times* after the leaking of confidential minutes of a meeting at which the energy minister Senator Parer revealed the Cabinet decision to oil and coal industry lobbyists. The minutes record Barry Jones of the Australian Petroleum Producers and Exploration Association saying: 'We can have that in writing? ... That is a resoundingly positive statement'. Parer also told the meeting that officers of the Australian Greenhouse Office had warned him that Australia's special deal had signalled to industry that it could 'sit back and do nothing'.[6]

The decision followed close on a visit to Australia by President Clinton's secretary of state, Madeleine Albright, who devoted her one major public address to urging Australia to implement the climate change treaty,[7] an intervention that prompted the aluminium lobbyist David Coutts to write in a private fax that Albright's speech 'reads like it was written by Al Gore and has caused some offence to the Government'.[8]

In August 2000 the Government announced that it had agreed on a set of commitments to Australian industry about the future of greenhouse gas abatement measures. Essentially, it was an announcement that set out to say what the Government would *not* do. It undertook: to minimise the burden on industry; not to introduce emissions trading; to involve industry in all phases of developing abatement policies; to avoid measures that discriminated against early movers; and to shun policies that distorted investment decisions.[9]

The last commitment was especially silly, as the only way to reduce greenhouse gas emissions over the longer term is to 'distort' investment decisions away from emissions-intensive industries and activities and towards low- and zero-emissions activities. The use of the word 'distortion' revealed that the Government continued to regard the existing energy economy as the natural state of affairs.

After the negotiations to refine the Kyoto Protocol in Marrakech in November 2001, the Government commissioned new modelling of the expected economic impacts of Australian ratification. The modelling was conducted not by ABARE, which had lost a great deal of credibility, but by one of the Government's favourite economists, the Australian National University's Warwick McKibbin.[10] McKibbin, who had developed his own, widely ignored global scheme to manage climate change, was and remains a vocal critic of the Kyoto Protocol. Nevertheless, his modelling concluded that the economic cost of the Kyoto Protocol would be higher if Australia did *not* ratify the treaty than if it did. It concluded that by 2010, compared with business as usual, Australia's gross national product (GNP) would decline by 0.4 per cent if Australia stayed out of the Kyoto Protocol, but would decline by only 0.33 per cent if Australia ratified. This is because actions by other countries (such as Japan reducing its coal imports) would have a negative economic effect, which Australia could partially offset if it started to cut its emissions too.

By 2020, however, McKibbin estimated that Australia's participation in the Kyoto Protocol would reduce real GNP by 0.51 per cent, compared with a fall of 0.30 per cent if Australia refused to ratify. How painful would it be to see our real GNP reduced by 0.21 per cent? Under business as usual Australia's real GNP will almost exactly double on about 1 December 2020.[11] If we ratified the protocol then, with the existing policies, our GNP would not double until the end of January 2021, a delay of eight weeks. This eight-week wait to become twice as rich has been the basis for the repeated stories about the huge economic costs Australia would face.

It was no wonder, therefore, that the Government refused to release the results of the modelling for five months and then did so at six o'clock on a Friday night, minimising the chances of their being reported. If accurate, they demolished the foremost rationale for Australia's refusal to sign up to the treaty. In his media statement accompanying the release of the modelling, Hill's replacement as environment minister, David Kemp, distanced the Government from the new evidence, claiming the work it commissioned only addressed 'a limited set of the issues'.

On the same theme, the Government frequently claimed that ratifying the Kyoto Protocol would result in massive job losses. During the 2000 federal election, a Government blitz throughout regional Australia claimed Labor's decision to ratify would see regional economies gutted. It relied on modelling by Allen Consulting and commissioned by the Minerals Council of Australia (that is, the mining industry). 'Cuts to greenhouse gases will hit GDP and jobs' and 'Victoria facing huge job losses' were typical headlines.[12] 'Kyoto ends 50,000 jobs', announced Rockhampton's *Morning Bulletin*, going on to declare: 'The Central Queensland mining industry would face massive job losses if a Labor government came into power and signed the Kyoto Greenhouse Agreement, a Queensland Mining Council representative has warned'.[13]

The studies on which these claims were based ranged from the dubious to the ridiculous. The Allen Consulting report which the Government seized upon suffered from several simple but vital errors, the effect of which was to ramp up the estimates of job losses.[14] For example, the report attributed all claimed job losses to 'Australian compliance' with the Kyoto Protocol, yet most of its forecast job losses were due not to greenhouse measures in Australia but to decisions by governments overseas. In some regions the claimed job losses arose from a predicted sharp decline in agricultural output due to the imposition of a large tax on methane emissions from livestock (a 'belch tax'), a policy response that was fanciful. And the modelling completely ignored the two largest tranches of emission cuts in Australia that are available at no cost or very low cost, namely, accelerated energy efficiency and the end of land-clearing.

To arrive at its assessments, Allen Consulting used a blunt 'carbon tax' policy which it knew from its previous work would be much more expensive than a mix of policies. It had conceded that a 'policy mix' was 'more realistic' and would impose minimal costs on the economy, yet it unaccountably opted for a much more expensive economic instrument. Coincidentally, every one of the errors and misinterpretations in the Allen Consulting report had the effect of exaggerating the apparent costs to GDP and employment of meeting Australia's Kyoto obligations.

The Government has always tried to have it both ways. It wouldn't ratify Kyoto, but it would meet the target. It repudiated the agreement, yet it wanted to keep its seat at the table. When it heard that some businesses were worried that they would be harmed by exclusion from Kyoto, it insisted that Australian firms could still participate in international emissions trading even if Australia did not ratify the Kyoto Protocol. At best this was wishful thinking; at worst the Government was misleading Australian businesses. The European Emission Trading System explicitly allowed the scheme to be linked up with trading schemes in other countries, but only if they had ratified the Kyoto Protocol. This was written in to the system. Because of the misleading statements emanating from various Australian ministers, in 2002 the European Commission's Delegation to Australia issued an unambiguous denial:

> On the question of carbon emissions trading, the Kyoto Protocol clearly states that carbon trading is allowed between

those Parties who have ratified the Protocol. Countries that are not Parties to the Kyoto Protocol are not eligible to participate in emissions trading under it. Nor can emission reduction projects or carbon sequestration efforts taking place in its territory be rewarded under the Protocol.[15]

EMISSIONS TRADING

The Kyoto Protocol had included provision for an emissions trading system among industrialised countries, one that developing countries could participate in through the Clean Development Mechanism. Under the trading system expected to operate during the 2008 to 2012 commitment period, nations would be able to sell credits if their actual emissions were less than their 'assigned amounts', thereby providing an incentive for some nations (and by extension major polluters within those nations) to cut emissions by more than the amounts agreed at Kyoto.

The attraction of emissions trading for environmentalists is that by greatly reducing the costs of greenhouse gas mitigation, more rapid reduction in global emissions should be permitted. Against this, the details of the trading system may open up a number of loopholes that relieve obligations on countries to reduce their output by the agreed amounts. There is a real possibility that trading will permit a flood of surplus emissions onto the world market that will significantly reduce pressures to find cheaper ways to cut emissions. Through this form of 'offshore compliance' the United States and Japan could simply buy their way out of much of the domestic cuts that their own targets seem to demand.

Even so, a trading system without loopholes offers the best opportunity of achieving sustained reductions because it minimises the cost of cutting emissions while maintaining the environmental integrity of the protocol.

In Australia, auctioning emission permits to polluters would provide a large new source of revenue for the Federal Government. Combined with 'revenue recycling' – that is, using this new revenue to reduce other taxes – emissions trading opens up the opportunity for a significant change to the tax system, one that could allow Australia to meet its obligations in the first and subsequent commitment periods and improve overall economic welfare at the same time. In addition to its ability to raise revenue and redistribute income, the taxation system can be a powerful device for changing behaviour. Many aspects of the existing tax system *discourage* activities that are socially beneficial – such as employment and investment – and *encourage* activities that cause environmental damage – including pollution and excessive use of resources. There is therefore scope to restructure tax systems in revenue-neutral ways that both promote greater employment and inhibit environmental damage. This is the idea behind ecological tax reform, an idea that was extensively discussed in the 1990s and was implemented in some European countries.

In 1999 the Australian Greenhouse Office commissioned a series of four discussion papers examining various aspects of emissions trading.[16] While not making any recommendations, the tenor of the papers was supportive of the introduction of trading in Australia before the initiation of international trading in 2008. Predictably, the fossil-fuel industries became alarmed at the move towards effective measures to tackle greenhouse gas emissions and began lobbying key ministers. The Government caved in and in August 2000 announced that it would not introduce a domestic trading system in advance of an international one.

In his November 1997 greenhouse policy speech, the Prime Minister had undertaken to introduce measures to require the electricity sector to source an additional 2 per cent of electricity from new renewable energy sources. The AGO proposed the Mandatory

Renewable Energy Target (MRET) scheme, which was immediately seen as a threat by the fossil-fuel lobby. However, a prime minister's unambiguous promise is difficult (although not impossible) to avoid and eventually a bill was drafted and legislation was introduced to parliament in early 2000. Despite the fact that the proposal had been watered down considerably after intense lobbying by industry, the renewable energy industries decided to support the bill as the delay had been holding up large amounts of investment. The final legislation capped the mandated amount of new renewables at 9500 gigawatt-hours (GWh) of electricity. With expected growth in electricity demand, this meant that Howard's 2 per cent promise ended up as a less-than-1-per-cent policy.

Nevertheless, even a small incentive like this one, since it was mandatory, led to an extraordinary boost to investment. While many had predicted that a large part of the 9500 GWh of new renewables would be accounted for by the burning of bagasse (sugar cane residue) in furnaces in sugar mills, it was the wind industry that responded most strongly to the incentives. In fact, we know from the leaked minutes of the LETAG meeting in May 2004 that the Government regarded the response of the wind industry as a threat to the coal-fired generators.

THE RETREAT TO VOLUNTARISM

As we have seen, the Australian Government's approach was based almost wholly on voluntary programs (with the notable exception of the MRET scheme, which was soon killed off). Rather than initiating and enforcing mandatory programs, the Federal Government offered a wink and a nod to the big polluters – both parties agreeing to give the impression that they were taking climate change seriously.

The Government was also enamoured of green consumerism. Greenpower schemes, information campaigns and programs such as Cool Communities, which demonstrated to households how they could reduce their energy use, received enthusiastic endorsement and funding. Every time a well-meaning environment group urged us to take responsibility for our own greenhouse emissions, the Government cheered because it immediately shifted responsibility away from itself. For all of its good intentions, green consumerism contributes to the progressive privatisation of responsibility for environmental degradation. Instead of it being understood as something requiring systemic reform of our economic and social structures, we are told that we each have to take responsibility for our personal contribution to the problem. The assignment of individual responsibility is consistent with the economic rationalist view of the world, which wants everything left to the unfettered market, even when the market manifestly fails.

Tim Flannery's book *The weather makers* has made a major contribution to public understanding of the dangers posed by global warming and has opened the eyes of many readers previously unaware of the magnitude of the dangers we face. However, it falls into the trap of green consumerism. After an eloquent statement of the implications of unchecked climate change, drawing on the work of hundreds of climate scientists, Flannery concludes his book by arguing that voluntary action by well-meaning consumers is the only way to save the planet:

> It is my firm belief that all the efforts of government and industry will come to naught unless the good citizen and consumer takes the initiative, and in tackling climate change the consumer is in a most fortunate position ... there is no need to wait for government to act.[17]

This is music to the Federal Government's ears and perhaps explains why the former environment minister Ian Campbell declared: 'I know Tim very well ... I've got a lot of respect for him ... he's a very good scientist'.[18] Yet it is a reckless conclusion to reach. We did not eliminate the production of ozone-depleting substances by relying on the good sense of consumers in buying CFC-free fridges. We insisted our governments negotiate an international treaty that banned them. We did not invite car buyers to pay more to install catalytic converters, the greatest factor in reducing urban air pollution. We called on our governments to legislate to require all car-makers to include them. Yet Flannery urges each of us to take responsibility in the belief that our responses to these noble appeals will transform the market: 'If enough of us buy green power, solar panels, solar hot water systems and hybrid vehicles, the cost of these items will plummet', and we will solve climate change in spite of the Government.[19]

As the US analyst Michael Maniates has written: 'A privatization and individualization of responsibility for environmental problems shifts blame from state elites and powerful producer groups to more amorphous culprits like "human nature" or "all of us"'.[20] The environment becomes depoliticised so that the major parties can share a common vision without getting into a potentially damaging bidding war over who will better look after the environment. In the end green consumerism has more in common with the conventional economist's belief in consumer sovereignty and individualism, one promoted by the ideologues of the right-wing think-tanks who argue that the way to solve environmental problems is to give consumers a choice. If consumers do not make green choices, then it is obvious that they prefer to live in a polluted and climatically transformed world. While well-intended individual actions are not to be criticised in themselves, when they are sold as the answer to environmental decline they actually obstruct the path to the real solutions. Some environmentalists who lead radically simplified lifestyles also contribute to the process of individualisation when they project a holier-than-thou attitude. As most of us will never choose to live as they do, their exhortations mean that our problems will never be solved.

Another manifestation of the dangers of individualisation might be called the rich man's indulgence syndrome. Various wealthy individuals have 'discovered' climate change and decided to devote some of their time and money to solving it. Such philanthropy (if that is what it is) is very welcome if it supports effective political activity aimed at arousing the public to protest that their governments are doing nothing, but it often morphs into a business opportunity. Alan Finkel, an Australian scientific entrepreneur who made a fortune in the United States, drives an electric scooter around Melbourne and has bankrolled a new environmental magazine called G, billed as 'a practical guide for treading lightly'.[21] To reward his editor-in-chief and himself for their efforts, Finkel has paid half a million dollars for two seats on the Virgin Galactic passenger flight into space in 2009. It is difficult to think of any single act that would generate more unnecessary greenhouse gases than sending a tourist into space. Virgin Galactic is the pet project of airline boss Richard Branson, another tycoon who cares deeply about climate change.

9. BUSINESS REALIGNMENTS

CRACKS IN THE EDIFICE

Within months of the Kyoto conference, the previously united front of Australian industry began to show signs of fragmenting. The natural gas industry had unaccountably maintained a position of public solidarity with the other fossil-fuel industries, even though it was apparent to everyone that natural gas, with much lower greenhouse gas emissions than coal and oil, would be the big winner from any move to cut greenhouse pollution. In 1999 it was reported that the Australian Gas Association, under its new executive director Bill Nagle, had broken away from the main fossil-fuel lobby group, the Australian Industry Greenhouse Network.[1]

The break caused considerable anxiety among some executives of gas companies, especially those whose interests overlapped with the coal and oil firms. But commercial imperatives won out. The Australian Gas Association released a report in February 2000 confirming that electricity from gas-fired generators produces only half of the greenhouse gas emissions of coal-fired electricity.[2] The report showed that, using best-practice technology in each case, greenhouse gas emissions from electricity fuelled by black coal are 67 per cent higher than emissions from gas generation, while emissions from brown-coal generation are 130 per cent higher.

One of the biggest gas companies, AGL, went further. In a submission to a parliamentary inquiry in September 2000 it argued that: 'The Kyoto Protocol represents a critical first step in the process of reducing global greenhouse gas emissions and, as a wealthy nation with one of the highest per capita emissions, Australia has a responsibility to take a leadership role internationally'.[3] AGL criticised the approval of new coal-fired power plants in Queensland, called on the Government to ratify the Kyoto Protocol without delay, and supported the rapid introduction of a domestic emissions trading scheme.

The splits were more complicated in the electricity industry. The Electricity Supply Association of Australia (ESAA) gave the appearance of public concern about climate change but in practice did all it could to block federal and state measures. Some of the more progressive electricity generators, such as Stanwell and Pacific Power, were uncomfortable with this, and the renewable energy suppliers, notably the Tasmanian Hydro Electric Corporation and the operators of the Snowy Mountain Scheme, began to argue that the industry should accept the need to cut emissions. The ESAA found it increasingly difficult to satisfy all of its members, and an organisation representing the brown-coal generators was formed in order to push its own hard-line stance. The Latrobe Valley Generators group comprised five privatised brown-coal Victorian generators, some owned by overseas corporations. It employed Noel Bushnell of Hannagan Bushnell to speak on its behalf. Hannagan Bushnell had been responsible for the media strategy behind the Countdown to Kyoto conference in Canberra in August 1997 and had acted as media strategists for greenhouse gas polluters at international conferences. In contrast to the words used by the Australian Industry Greenhouse Network, which declared (in public at least) that it would support ratification of the protocol if certain conditions were met, the Latrobe Valley Generators opposed ratification outright, challenged the science of climate change and used extravagant language about threats to

Australian sovereignty similar to that employed by the sceptics on the far right.[4]

Before Kyoto, the sustainable energy industries were weak and poorly organised, but the commercial stimulus to low- and zero-emissions technologies provided by the agreement gave those industries an organisational fillip. The Sustainable Energy Industry Association relaunched itself as a more professional industry lobby, throwing off its old image of a collective of fringe industries for enthusiasts only. In 2002 it merged with the Australian Eco-Generation Association to become the Australian Business Council for Sustainable Energy, an organisation that has gradually built itself into a stronger and more coherent voice for low-emissions industries.

The position of the Business Council of Australia, representing the interests of Australia's biggest 100 corporations, was also changing. With Hugh Morgan playing the general's role, the BCA had consistently opposed – often in the crudest way – any talk in Government of implementing Australia's Kyoto obligations. Most BCA members had run dead on the climate issue, leaving the field to the mining or fossil-fuel companies. However, in late 2000 the ranks of the BCA were split by the decision of BHP, under its new American managing director Paul Anderson, to take a more progressive stance. BHP released a proposal for implementing 'credit for early action' based on the introduction of a domestic emissions trading system in advance of the international one. The move caused severe ructions.

In late 2002 a serious push began for the BCA to switch from opposing to supporting ratification of the Kyoto Protocol. Backed strongly by BP's Greg Bourne and Westpac's David Morgan, the move was well organised and resulted in a memorandum from the head of the BCA's greenhouse task force, Meredith Hellicar, arguing that the Government's position was out of step with the community and that the cost to the nation would be 'greater if we do not ratify'.[5] The move provoked a bitter backlash from the conservative energy companies.[6] A number of strongly worded letters from the fossil-fuel giants were immediately dispatched to Hellicar and BCA executive director Katie Lahey. The contents of the letters were so similar that they were clearly orchestrated, almost certainly by Hugh Morgan.[7] A number contained veiled threats about leaving the BCA. Wayne Osborn of Alcoa wrote that his company 'simply could not support the approach taken in this memo in any form'. Woodside declared it 'entirely inappropriate' and the chair of Exxon, Robert Olson, declared it would 'cause a split in the BCA membership'. They all referred to 'credible studies' that indicated serious economic damage would be done and claimed that the shift would incur the wrath of the Federal Government.

The stakes were high, for if the nation's peak business group backed ratification, the Howard Government would no longer be able to claim that its refusal to join Kyoto was in the interests of the economy. It would be severely embarrassed at home and abroad. The lobbying was intense, and after a series of meetings in February 2003, the BCA announced that its members could not agree and that the council 'is not in a position at this time to either support or reject ratification'. It was half a victory but left the resource companies and the Government a great deal of room to continue on the course they had set. To buttress their position, the resource companies backed Hugh Morgan to become president of the BCA in 2003, a position he held for the next two years.

THE ALUMINIUM INDUSTRY

From the mid-1990s, the loudest cries about how cutting Australia's emissions would damage 'competitiveness' came from the aluminium industry. To protect its interests, the industry organised a low-profile but highly effective lobby group, the Australian Aluminium Council (AAC), which operates out of unprepossessing offices in suburban Dickson in Canberra. The AAC exists almost solely to lobby against the introduction of measures to combat climate change. Over 95 per cent of its many media releases over the years have dealt with this topic. The moving force behind the organisation was Hugh Morgan, whose company, Western Mining Corporation (WMC), had a major stake in Australia's second-largest aluminium smelter at Portland.[8]

As a key member of the Australian Industry Greenhouse Network, and thus the greenhouse mafia, the Aluminium Council has been at the forefront of industry claims that mandatory targets would cause severe economic damage in Australia.[9] It has argued that the burden for cutting emissions should be placed on other sectors of the economy and households rather than being distributed equally across the economy. Its constant refrain is that measures to restrict emissions would damage its international competitiveness, resulting in lost market share and a decline in Australian economic welfare. Its biggest weapon has been the threat to take its investment offshore, a threat it issues to governments around the world in order to secure public subsidies and avoid regulation.

The aluminium industry was one of the business groups to contribute $50,000 to gain a place on the ABARE steering committee. Aluminium companies were also some of the largest sponsors of the Countdown to Kyoto conference.

In the late 1990s the aluminium-smelting industry accounted for 16 per cent of greenhouse gas emissions from the electricity sector and 6.5 per cent of Australia's total emissions.[10] Australia has only six smelters, three large ones at Boyne Island, Tomago and Portland and three smaller ones at Kurri Kurri, Point Henry and Bell Bay. Of their total aluminium output 79 per cent is exported. These exports were worth around $2.8 billion in the late 1990s, when the smelting industry employed around 5350 people. Overall, foreign-owned companies account for around 59 per cent of the output of the aluminium-smelting industry in Australia, with Japanese (17 per cent), UK (14 per cent) and US (12 per cent) interests dominant. The level of control is substantially higher.

Aluminium is sometimes referred to as 'congealed electricity' and the smelters have extracted heavily subsidised electricity prices from state governments keen to attract major industry. The prices paid for electricity by aluminium smelters are set in long-term contracts and are a closely kept secret. However, enough information is available to make a good estimate of the extent of subsidies. The general belief in the electricity industry is that smelters pay between 1.5 and 2.5 cents per kilowatt-hour for delivered electricity. This can be compared with the cost of around 5 or 6 cents paid by other large industrial users, and 12 cents or more paid by households.

In 1997, the treasurer in the Victorian Kennett government, Alan Stockdale, complained that other high-voltage customers were paying up to three times the price paid by the two Victorian smelters.[11] The Victorian Department of Treasury and Finance described the contracts to supply Portland and Point Henry with electricity from the Loy Yang B power plant as 'onerous and unfavourable', and indicated in 1997 that they were 'costing the Government over $200 million per year'.[12] These two smelters account for one-third of total output in Australia. Overall, the total financial subsidy to aluminium smelters in Australia is estimated to be $410 million per annum.[13] In addition, the aluminium-smelting industry is responsible for a large proportion of Australia's greenhouse gas pollution – a cost imposed on others which can be valued by the anticipated cost of permits to emit greenhouse gases. The industry has said that it believes it should not be required to pay the costs of its pollution, and that other sectors of the economy should bear the entire burden. It is estimated that the failure to pay for the costs of aluminium-smelting pollution amounts to an additional subsidy to the industry worth at least $430 million per year.[14]

If the aluminium smelters carried through with their threat to shift out of Australia in response to the introduction of greenhouse gas abatement policies, their departure would result in a net economic benefit to Australia. Every dollar of income from primary aluminium exports costs more than a dollar to produce when public subsidies are accounted for. Through industry development programs and wage subsidies, the annual $410 million in direct financial subsidies freed up could be used to provide many more jobs than the aluminium industry currently provides. Indeed, each and every one of the industry's employees could be paid $70,000 to stay at home and there would still be funds left over. In addition, by saving 28.5 million tonnes in greenhouse gas emissions per year the departure of the industry would make it a great deal easier for Australia to meet its Kyoto target, by freeing up a large tranche of emissions for other, unsubsidised sectors. Yet the industry has the Howard Government in its thrall. Again and again it has bluffed and bullied governments into giving it special concessions and has consistently retarded progress on the greenhouse issue. It has been the most self-serving and uncompromising lobby group in the climate change debate in Australia.

This helps to put into perspective one of the favourite arguments used by the Government to refuse to ratify the Kyoto Protocol, the claim that Australian firms would shift offshore, which is advanced with aluminium smelters foremost in mind. It has to be asked why an aluminium company would shift a smelter with a 30- to 40-year lifespan to a developing country to escape greenhouse restrictions in Australia, when everyone accepts that developing countries will also have to take on emission-reduction obligations within a decade or so. They must have very short-sighted CEOs. The hollowness of the threats was revealed in the late 1990s when the industry committed $3 billion to build a brand-new smelter and refinery at Gladstone in Queensland with a lifespan of four decades.

The Government also argues that any shift of industry offshore would be worse for the environment.[15] Apart from recycling the erroneous belief that all developing countries are dirty, polluted and inefficient, in the case of aluminium the fact is that Australian smelters produce more greenhouse gases per tonne of aluminium than smelters anywhere else in the world. Australian smelters' emissions from electricity consumption are 13.6 tonnes of carbon dioxide per tonne of aluminium, around 2.5 times the world average.[16] They are so high because Australian smelters rely almost wholly on electricity from coal-burning. According to the International Aluminium Institute, smelters in developing countries are cleaner than those in developed countries, producing lower direct greenhouse gas emissions per unit of output.[17] So if Australian smelters shifted anywhere else, global greenhouse gas emissions would fall. Besides, respectable corporations nowadays don't threaten to take their dirty factories to poor countries so they can exploit lax environmental laws. Yet that is how the Federal Government seems to view the matter.

In addition, it is a questionable assumption that Australia uses energy more efficiently than those developing countries to which industry might shift. The Government has argued that Australia should be permitted to become a global specialist in fossil fuels because we are much more efficient users of them than Asian countries. It is true that current electricity production from coal in Australia is between 15 and 25 per cent more efficient (in use of coal and thus carbon dioxide emissions) than installed capacity in typical Asian countries. However, many of the relevant developing countries are rapidly catching up, and any new facilities are likely to use globally available state-of-the-art technology.[18] Current relative efficiencies are no guide to future relative efficiencies.

Moreover, it is now broadly accepted that increasing electricity demand in Australia is more likely to be met by refurbishment of existing coal-fired power stations – some built in the 1970s – than by greenfield coal-fired power stations. The privatised Hazelwood power station in the Latrobe Valley had its life extended by 30 years in the late 1990s. Yet the thermal efficiency of Hazelwood is only 24 per cent, less efficient than many existing coal-fired stations in developing countries, and well below the 35 per cent that would be expected from a new greenfield station.

While the Australian Aluminium Council has mobilised more anti-Kyoto lobbying power than any other industry group, the parent companies of the biggest smelters in Australia – including Alcoa and Rio Tinto – have signed up to the United States–based Pew Center's Business Environmental Leadership Council, which favours implementing the Kyoto Protocol. Yet the Australian subsidiaries of these companies have been the most powerful forces in Australia's blocking action on climate change. Alcoa and Rio Tinto stand out for their hypocrisy. In Australia, Rio Tinto engages in expensive greenwash campaigns to convince the public of its environmental credentials. The industry often produces figures designed to show that it has reduced its greenhouse gas emissions, but these always apply only to the direct emissions that occur on the smelter sites and never to the emissions from the coal-fired power plants from which they draw their electricity.

But in the eyes of the Howard Government, these companies must be looked after at all costs. As I will discuss in the next chapter, this unseemly alliance is built on a network of personal connections. The moving force behind the Australian Aluminium Council, Hugh Morgan, is a close personal friend of the Prime Minister, at least since Howard appointed him to the Reserve Bank Board in 1981. They talk regularly. The successive executive directors of the Aluminium Council, David Coutts and Ron Knapp, previously served at senior levels in the Canberra bureaucracy and continue to cultivate their personal links. And to cap it off, in 2002, after years in the diplomatic service, the head of the Australian delegation to the Kyoto conference, Meg McDonald, joined Alcoa as general manager of corporate affairs in Australia. As Australia's position at Kyoto had been effectively dictated by the greenhouse mafia, McDonald did not need to make any adjustments to her stance on climate change on taking up her new job.[†]

DEBATING THE COSTS

While industry groupings were fragmenting and growing numbers of big businesses abandoning their oppositionist stance, the battle of the economic models continued. In a report timed to appear just before The Hague conference, Allen Consulting claimed that meeting the Kyoto target would cost $44 for every tonne of carbon dioxide emissions reduced, a big number quoted enthusiastically by the Latrobe Valley Generators. This was the headline figure from a report, commissioned by the Kennett Government and the Minerals Council, that actually contained four policy scenarios.[19] One of them factored in the opportunities for costless cuts from energy efficiency gains and reductions in land-clearing and concluded that the cost of meeting the Kyoto targets would be almost nil. Needless to say, this scenario – far and away the most likely one, for why would the Government choose much more costly policies? – was not promoted and received almost no attention.

[†] In 2006, McDonald was named president of the New York–based Alcoa Foundation, a company fund designed to enhance its corporate image by providing grants to various programs, assisting charities and so on. On her appointment, Alcoa's chairman and CEO said: 'Her expertise in negotiating key global issues in the United Nations and other international organisations will be a valuable asset ... as Alcoa expands its presence around the world' (Alcoa, *New Alcoa leadership*, media release, Pittsburgh, 15 February 2006, <http://www.alcoa.com/global/en/news/news_detail.asp?pageID=20060215005734en&newsYear=2006>).

Like other economy-wide modelling exercises, Allen Consulting's analysis assumed that firms act fully rationally – in the sense that they already do all that they can to maximise profits. In particular, it assumed there were no energy efficiencies to be had. This is an article of faith of neo-classical economists, rather than an assessment based on practical observation. Several studies have shown that energy consumption in Australia can be cut by between 20 and 40 per cent at no net cost.

If the Government had been inclined to believe the abatement cost figure of $44 per tonne, it would have been embarrassed by another, more believable set of figures emerging from its own Australian Greenhouse Office. Under the goods and services tax deal struck with the Democrats, the Government had allocated $100 million a year for four years to the Greenhouse Gas Abatement Program. The program was aimed at commercial applicants who would commit to reduce emissions by at least a quarter of a million tonnes per year. In the second half of 2000 the AGO received over 100 applications from firms around Australia. A number of good applications sought funding of around $3–4 per tonne or less. Predictions of economic ruin based on modelled abatement costs of $44 a tonne of carbon dioxide seemed absurd when placed against real bids by commercial decision-makers of $3–4 a tonne.†

The Federal Government continued to argue that cutting emissions in Australia would be more expensive than in other OECD countries. But if cutting emissions was so high in Australia, this left unanswered the question of why Japanese companies were coming to Australia to buy emission credits (albeit credits attached to tree plantations).[20] While the Minerals Council spent big money on economic

---

† Although it should be borne in mind that as the requirement to cut emissions increases so does the unit cost. Thus the estimate of $44 a tonne is the cost at the margin for a given economy-wide target, while the lower figures are 'inframarginal' costs, that is, below the cost at the margin.

models to prove that abatement costs were higher in Australia than elsewhere, electricity corporations in Japan were spending big money investing in Australia's cheap abatement opportunities.

August 2002 saw the publication of a statement signed by more than 250 of Australia's academic economists calling on Prime Minister John Howard to ratify the Kyoto Protocol without delay. Among the 254 signatories were 39 professors of economics, many of them quite conservative in their political views and economic orientation. The statement directly challenged the Government's rationale for refusing to ratify Kyoto on economic grounds: 'As economists, we believe that global climate change carries with it serious environmental, economic and social risks and that preventive steps are justified'.

The Government studiously ignored the views of the economics profession, preferring to hear only that advice which confirmed its predudices. There were moves to organise a 'counter-petition' by a right-wing economist at the Australian National University, Alex Robson, and backed by Alan Moran from the Institute of Public Affairs.[21] Their statement cast doubt on the science of climate change, referring to the 'uncertain science' and the absence of 'conclusive scientific evidence regarding the effect of human activity on global warming'. It criticised the proposals in the statement signed by the 254 economists as 'ill-conceived, poorly chosen policy responses'. The statement was posted on an anti-greenhouse libertarian website with links to the Centre for Independent Studies,[22] and received the backing of Piers Akerman, who, because he has no qualifications in the area, could not sign it himself.[23] Predictably, the Government's favoured anti-Kyoto economist, Warwick McKibbin, lent enthusiastic support to this far-right gaggle – the statement had, after all, praised his own scheme as a 'serious alternative' to Kyoto. The counter-petition met with indifference from the economics profession, and, with no support to speak of, it sank without a trace.

THE CHIEF SCIENTIST

In June 1999, the Prime Minister's Science, Engineering and Innovation Council, which draws together high-powered business interests and eminent scientists, issued a report urging the Government to go from defence to attack on climate change policy. The report described the Kyoto Protocol as a watershed in the global greenhouse debate and argued that it was a powerful instrument of change that would be ignored at great cost. It drew an analogy with earlier industrial and social movements:

> In each, attitudes changed from defence and denial, to recognition of opportunities, and ultimately to the realisation that what is right for the community in the long term can be good for the growth and profits of industry ... Increasingly the world's major corporations accept this transition ... The working group believes ... that those industries and countries which choose to do nothing may well in the long run be seriously disadvantaged ... If we wait for ratification while other countries act, Australia runs the risk of missing out on global opportunities, and may be left behind in terms of greenhouse compliance.[24]

The report was prophetic but it was not the sage advice the Government wanted to hear. It moved to ensure that it would not be embarrassed again and changed the membership of the group. In that same year the Prime Minister appointed Robin Batterham as chief scientist and chair of the Innovation Council, advising on national research priorities and science, energy and innovation policies. At the time Batterham was chief technologist at Rio Tinto, a position he retained while serving as chief scientist of the Commonwealth. Batterham had previously been employed by Comalco as its head of research and technology support. His position with the Commonwealth was for two days a week, and paid about $90,000 per annum; from Rio Tinto he received about $250,000 per annum.

Batterham was a strong supporter of geosequestration – also known as carbon capture and storage – as one of the keys to solving the greenhouse problem, something with which he had considerable familiarity as Rio is a major coalmining company which also has interests in energy-intensive industries. When asked by *Lateline*'s Tony Jones whether he supported the Government's repudiation of Kyoto, the chief scientist responded: 'That's correct, and this is one, I think, where we need some deep thought'. He went on to provide his deep thoughts.

> Kyoto agreement is what I would call a baby step in the right direction, and it comes down to the question of saying, 'Is a baby step in the right direction affecting the part of the world that is not going to be the main energy consumers over the next 30 years, because that's what all the projections suggest? Is this baby step in fact deluding us into thinking that we've got the problem solved?' So I come out and say, 'No, no, no', we've actually got to look at how we learn to not just take baby steps, but to walk, to run and, in fact, to sprint towards the line of 50 per cent reductions by 2050.[25]

In 2004, unpersuaded by Batterham's claims that he had no conflict of interest between his two roles, Labor and the Greens initiated a parliamentary inquiry to investigate the matter.[26] The following exchange between Senator Bob Brown and Dr Batterham suggests that the chief scientist had an enviable capacity to compartmentalise his working life:

SENATOR BROWN: Carbon capture, as you put it, is one of the central matters to be taken into account in achieving this desired outcome. What is the interest then for Rio Tinto in carbon capture and, in particular, in geosequestration? Putting it directly again: this is of central interest to Rio Tinto.
DR BATTERHAM: Firstly, I am not a spokesperson for Rio Tinto; that is not my job with them. I am an adviser to Rio Tinto as well on threats and opportunities. I am sure that you can get from the energy group in Rio Tinto a comment on how they see geosequestration. I am simply saying – and I can say it quite consistently and have done so for some years now – that carbon capture and storage is one of the four principal routes that we have got to follow.

The Senate inquiry found that the chief scientist had a 'clear conflict of public and private duties' arising from his dual part-time roles.[27] It also found that he used 'unpublished and unverified data supplied by Rio Tinto in a meeting of Commonwealth and state energy ministers, and failed to declare the source of the information', which contributed to a perception of a conflict of interest. Bob Brown called for his immediate resignation.[28] Under intense pressure, Batterham did not seek a third term as chief scientist and he now works full-time for Rio Tinto.

10. THE DENIALISTS

DOUBT IS OUR PRODUCT

The political battle over climate change has now been raging for 15 years. Throughout the debate the constant has been the ever-accumulating body of scientific evidence. Each new report has confirmed fears and rung the alarm bells more loudly. The Fourth Assessment Report of the IPCC in 2007 added to the accumulation of evidence and sharply reduced levels of uncertainty about the causes and impacts of global warming. But by itself a mass of scientific evidence, no matter how compelling, can have no effect; it must influence the views of decision-makers, and this means governments. Pressure can be applied directly, or through those to whom governments listen. In democracies, enormous efforts are made to influence public opinion and this fact gives the media a vital mediating role. Governments also listen to business because they believe that it is the creator of wealth and thus national well-being. And business is the principal source of funding for conservative political parties.

These links in the decision-making chain have provided an opportunity for political players to intervene to influence public debate over climate change. As climate change is the global environmental issue *par excellence*, it is no surprise to learn that the arguments and tactics used by the denialists and the greenhouse mafia have international currency. Most have emanated from the United

States, and it is surprising to discover the extent of the subterranean influence on climate change politics in Australia exerted by shadowy right-wing organisations in the US.

An analogy is sometimes drawn between those who have resisted the tide of scientific evidence on the dangers of climate change and those who once questioned the link between smoking and lung cancer in the face of overwhelming medical evidence. It turns out that the links between denialists in the climate change and smoking controversies go much deeper than mere analogy. In response to the 1992 report of the US Environmental Protection Agency linking passive smoking with cancer, Philip Morris hired a public relations company named APCO to develop a counter-strategy.[1] Acknowledging that the views of tobacco companies lacked credibility, APCO proposed a strategy of 'astroturfing', the formation and funding of apparently independent front groups to give the impression of a popular movement opposed to 'overregulation' and in support of individual freedom. Foremost among the fake citizens' groups was The Advancement of Sound Science Coalition (TASSC). According to secret documents, it was to be 'a national coalition intended to educate the media, public officials and the public about the dangers of "junk science" … Upon formation of Coalition, key leaders will begin media outreach, e.g. editorial board tours, opinion articles, and brief elected officials in selected states'.[2]

The strategy was to link concerns about passive smoking with a range of other popular concerns, including global warming, nuclear waste disposal and biotechnology, in order to suggest that these were all part of an unjustified social panic, and hence calls for government intervention in people's lives were unwarranted. It set out to cast doubt on the science, to link the scare against smoking with other 'unfounded fears' and to contrast the 'junk science' of their opponents with the 'sound science' that they promoted. As one tobacco-company memo noted: 'Doubt is our product since it is the best means of competing with the "body of fact" that exists in the mind of the general public. It is also the means of establishing a controversy'.

As the 1990s progressed, and the rear-guard action against restrictions on smoking faded, The Advancement of Sound Science Coalition started receiving funds from Exxon (among other oil companies) and its 'junk science' website began to carry material attacking climate change science. George Monbiot, who uncovered these links, wrote that this website 'has been the main entrepot for almost every kind of climate-change denial that has found its way into the mainstream press'. Having been set up by Philip Morris, TASSC 'was the first and most important of the corporate-funded organisations denying that climate change is taking place. It has done more damage to the campaign to halt it than any other body'. The anti-greenhouse messages that it has developed and disseminated have been taken up and used by denialists around the world, and particularly in Australia. They have been reproduced in dozens of opinion pieces and even news stories in newspapers around the country. Wrote Monbiot:

> By dominating the media debate on climate change during seven or eight critical years in which urgent international talks should have been taking place, by constantly seeding doubt about the science just as it should have been most persuasive, they have justified the money their sponsors have spent on them many times over.[3]

Virtually every argument and tactic used by the denialists in Australia has been imported from the United States. The occupation of the White House by George W. Bush – a feat achieved on the back of huge donations from the oil industry – gave the denialists a massive boost. The approach of the Bush Administration was set out in a leaked memo prepared in November 2002 by Frank Luntz, a Republican Party political consultant. In a section titled 'Winning the Global Warming Debate', Luntz wrote that the environment is the issue on which Republicans and President Bush were most vulnerable: 'Should the public come to believe that the scientific issues are settled, their views about global warming will change accordingly. Therefore you need to continue to make the lack of scientific certainty a primary issue'.[4]

One trusted method of undermining the credibility of climate science has been to exploit public ignorance of how science works. Thus it is often claimed that scientists in the 1970s were warning against a new ice age. 'And now they are telling us the Earth is warming!' George Monbiot observed: 'This is rather like saying that because Jean-Baptiste Lamarck's hypothesis on evolution once commanded scientific support and was later shown to be incorrect, then Charles Darwin must also be wrong'.[5] The fact that climatologists who in the 1970s predicted an ice age were shown to be wrong and the idea discarded is a cause for confidence in climatology, not distrust.

In Australia, the broad links between fossil-fuel companies, sceptics groups and media commentators, mainly in the Murdoch press, have been apparent even if the details have been kept secret. It is, nevertheless, remarkable to discover that the arguments, tactics and even the personnel used by tobacco companies were shifted across to the fossil-fuel corporations. In the climate change debate they have been used to much greater effect.

Among the other important denialist organisations funded by ExxonMobil has been the website Tech Central Station, which has links to Australian groups. It describes itself as a website 'where free markets meet technology' and is probably the world's most effective climate sceptic website. In addition to ExxonMobil, Tech Central Station receives funding from General Motors, McDonald's and drug companies. It publishes daily original commentary, news and analysis focused on economics, business, foreign affairs, technology, science, the environment, trade and culture.[6] Until recently the website was published by the DCI Group, a top Republican lobbying and public relations firm with close ties to the Bush Administration.[7] The DCI Group offers services that include: national, state and local lobbying; coalition building; and generating 'grasstops' and constituent support for issues. DCI advertises its ability to provide 'third party support' to clients and has been linked to several industry-funded coalitions that pose as grassroots organisations. 'Corporations seldom win alone', the group's website says.

> Whatever the issue, whatever the target – elected officials, regulators or public opinion – you need reliable third party allies to advocate your cause. We can help you recruit credible coalition partners and engage them for maximum impact. It's what we do best.

The company's skills in astroturfing were acquired by its managing partners – Tom Synhorst, Doug Goodyear and Tim Hyde – during nearly a decade of work in the 1990s for R.J. Reynolds Tobacco Company.[8]

In addition to front groups and industry-funded websites, a number of right-wing think-tanks have played a crucial role in preventing action on global warming. Perhaps the foremost has been the Competitive Enterprise Institute, a Washington-based conservative think-tank 'dedicated to advancing the principles of free enterprise and limited government'.[9] Among its many statements denying the seriousness of global warming are claims that climate change would create a 'milder, greener, more prosperous world' and that 'Kyoto was a power grab based on deception and fear'.[10] Corporate funders include the American Petroleum Institute,

Cigna Corporation, Dow Chemical, EBCO Corp, General Motors and IBM, as well as ExxonMobil.[11] The Competitive Enterprise Institute is also intimately involved in the Cooler Heads Coalition, which argues that the risks of global warming are speculative. Just before the release of Al Gore's film *An inconvenient truth* in 2006, the Competitive Enterprise Institute took out television advertisements arguing against climate change. Notoriously, one of the ads ended with the words: 'Carbon dioxide – they call it pollution, we call it life'.

These groups have spawned and emboldened a network of individuals who have little scientific training but who are utterly convinced that the 'global warming theory' is a giant fraud being committed by the scientific establishment. They crop up everywhere, but particularly on the internet. The Wikipedia entry on global warming, written by a respected climate modeller, was repeatedly changed by an anonymous sceptic.[12] This 'revert war' went on for a year before the managers of the online encyclopaedia placed a six-month ban on the sceptic and limited the ability of the climate scientist to engage in 'revert' entries. A resumption of hostilities is now expected with the release of the Fourth IPCC Assessment Report.

AUSTRALIAN CONNECTIONS

In Australia, two men have been at the centre of the campaign to prevent the Federal Government from taking action to cut emissions – Hugh Morgan and Ray Evans. As CEO of Western Mining Corporation (WMC) from 1990 to 2003, Hugh Morgan has been one of Australia's most influential business figures. His father built WMC into a major player and the young Hugh was employed in the company, cultivated and promoted by his father's successor as company head, Arvi Parbo.[13]

Morgan has also been an influential member of the Liberal Party, including serving as a director of the Cormack Foundation, the most important Liberal Party fund-raising vehicle.[14] He has co-chaired the foundation with John Calvert-Jones, the federal treasurer of the Liberal Party and the brother-in-law of Rupert Murdoch. The foundation raises money from some of the biggest fossil-fuel companies, including BHP, Wesfarmers and Rio Tinto, and in 2004–05 it was the single biggest donor to the Liberal Party, giving $1.9 million. Since 1998 it has donated a total of $10.64 million to the party.[15] WMC (taken over by BHP-Billiton in 2005) was also a major donor. In 2000–01, for example, it gave $240,000 to the Liberal Party, $50,000 to the National Party, and nothing to Labor. In 2001–02 WMC gave a further $170,000 to the Liberals. Morgan personally has donated $69,900 to the Liberal Party since 1998–99. Jeff Kennett says Morgan has had a 'profound influence' on Australian policy debates.[16]

Morgan has been at the centre of right-wing political activism in Australia since the 1980s, including close association with the neo-liberal H.R. Nicholls Society, founded by Ray Evans, Peter Costello and John Stone. He came to public prominence in 1984 when, as president of the Mining Industry Council, he made a savage attack on Aboriginal spirituality and land claims. He wrote that land rights represented a 'symbolic step back to the world of paganism, superstition, fear and darkness', that mining is next to godliness and that Aboriginal rights to land had been forfeited because they had practised 'infanticide, cannibalism and cruel initiation'.[17] Aboriginals, he claimed, had shown a preference for 'the particular flavour of the Chinese, who were killed and eaten in large numbers'.[18]

In recent years Morgan has been a frequent visitor and confidant of Prime Minister Howard, who first appointed him to the board of the Reserve Bank in 1981 (during Howard's time as Treasurer). In the early 1990s Morgan expressed equally crude, if not quite as insulting, views about environmentalism. He described ecology as a 'religious movement of the most primitive kind'.[19] He reserved particular fury for claims that the burning of fossil fuels is the cause of global warming, and in the late 1990s argued that the Kyoto Protocol would open Australia to foreign invasion. It is not known how often the Prime Minister has turned to Morgan for advice on greenhouse policy, but it is fair to assume that when Howard deferred a Cabinet decision on a carbon tax so he could consult with business leaders, Morgan's phone rang soon afterwards. According to one industry insider: 'In terms of who the Prime Minister listens to on the issue, he would definitely take notice of what captains of industry said to him – for example, Hugh Morgan and Barry Cusack [of Rio Tinto]'. Another insider noted that Morgan is progressive on some issues but 'on greenhouse, absolute luddite ... He thinks the world will fall apart under a Kyoto Protocol. So he's been quite influential'.[20]

Morgan's influence has been felt through other powerful groups. His company had major holdings in the Portland and Point Henry aluminium smelters and has been one of the leading members of the Australian Aluminium Council (others include Rio Tinto and Alcoa), which has been the most powerful anti-Kyoto business lobby. The Aluminium Council has in turn been a moving force within the Australian Industry Greenhouse Network, the organisational form of the greenhouse mafia, and Morgan has exercised his influence through that channel too. Between 2003 and 2005 Morgan served as president of the Business Council of Australia, where he worked hard to resist attempts to shift the organisation to a pro-Kyoto stance.

Morgan has a long history of greenie-hating and has for years been making feverish declarations of national catastrophe should environmentalists have their way. His two pet hates came together with the decision by the Hawke Government to ban mining at Coronation Hill in Kakadu National Park. In 1991 Morgan addressed the Adam Smith Club with a sense of foreboding:

This decision will undermine the moral basis of our legitimacy as a nation, and lead to such divisiveness as to bring about political paralysis ... The implications of it will, inevitably, permeate through the entire body politic, and cause, imperceptibly, like some cancerous intrusion, a terminal disability ... Like the fall of Singapore in 1942, Coronation Hill was a shocking defeat.[21]

To date, the mining ban at Coronation Hill has not led to the erosion of the moral basis of our legitimacy as a nation, or to political paralysis, or to a terminal disability. In the same speech, Morgan called for a counter-attack on the religious crazies and green antinomians 'who threaten our prosperity and eventually our survival'. Perhaps the Lavoisier Group, discussed below, was the long-awaited counter-attack.[22]

On 10 June 2002, *The Australian* published an opinion piece by Hugh Morgan in which he lavished praise on the Prime Minister for declaring unambiguously in parliament the previous week that his Government would not ratify the Kyoto Protocol. Using his customary hyperbole, Morgan characterised the protocol as a plot by European bureaucrats to centralise world power in Bonn, where the Kyoto secretariat would have 'wide-ranging powers of inspection and enforcement' and change international law to penalise any country that failed to accede to its demands. 'Australia would then be powerless to recover the sovereignty that had been de facto yielded up with ratification of Kyoto ... The Prime Minister has made the right call on Kyoto.'[23] Also in the papers on that same day was the announcement that the Government was making Morgan a Companion in the Order of Australia. As if to disprove the view that public servants are humourless, the citation said that the award was particularly for Morgan's 'leadership in the formation and evolution of sustainable development policy'.

On reflection, when we consider the anti-greenhouse activities in which Hugh Morgan has been intimately involved – the formation of the Lavoisier Group, his driving roles in the Minerals Council of Australia, the Australian Aluminium Council and the Australian Industry Greenhouse Network, the battles within the BCA and the lobbying and organising of WMC itself – it is clear that all of these had one purpose and one purpose alone: to apply pressure to his friend John Howard.

Ray Evans has always played Tonto to Hugh Morgan's Lone Ranger. An electrical engineer by training, he worked for the right faction of the ALP but split from it in the 1960s over the party's stance on the Vietnam War.[24] Evans was appointed as an executive officer at WMC in 1982 and served for many years as Morgan's loyal political operative, effectively acting as a one-man secretariat for some of the most important anti-greenhouse activities of Australian industry. He seemed to become more active in 1993 after the board of WMC rebuked Morgan for his relentless political activities and instructed him to devote himself to the management of the company. Of his relationship with Morgan, with whom he established an 'instant rapport',[25] Evans has said: 'My role was to engage in the culture wars and provide him with feedback'.[26] He left WMC in 2001, shortly before Morgan, to set up his own consulting company.

Evans is an office-holder in and apparent creator of a string of right-wing activist groups including the H.R. Nicholls Society (of which he is president), the Bennelong Society (serving as secretary), the Samuel Griffith Society (treasurer) and the Lavoisier Group. The Bennelong Society was established to promote a conservative approach to Aboriginal policy (and is linked to Keith Windschuttle's website). At one point it shared a post-office box with the Lavoisier Group.[27] Evans has described himself as 'a Genesis 1:28 man'.[28] The relevant passage from the Bible declares: 'God said to them "Be fruitful and become many and fill the Earth and subdue it, and have in subjection the fish of the seas and the flying creatures of the heavens and every living creature that is moving upon the Earth"'.

Evans was present at a seminal meeting held in November 1996 at the headquarters of the Exxon-funded Competitive Enterprise Institute in Washington. At this meeting a strategy was formulated to scuttle Kyoto. According to Australian journalist Bob Burton, other participants included a senior world vice-president for Ford Motors, American Petroleum Institute executive director Bill O'Keefe, and Dick Lawson, the executive director of the US National Mining Association.[29] R.J. Smith, senior environmental scholar with the Competitive Enterprise Institute, told Burton:

> Right after Tim Wirth of the U.S. State Department announced they were going to call for mandatory controls in Kyoto, we said, 'What do we do? How do we stop this?'
> It was clear that Australia if possible would be a key player in this, so we decided to see if we could plan a series of conferences before Kyoto and had the first one on July 15, 1997 in Washington, DC.

Among the speakers at the Washington conference were Brian Fisher, the head of ABARE[†], and Paul O'Sullivan, the Australian

---

† This was perhaps the start of Fisher's close relationship with the fossil-fuel lobby and right-wing anti-greenhouse organisations. He shared platforms around the world with some of the most zealous anti-greenhouse business groups and think-tanks, including Charles River Associates, a US consulting company that prepared alarmist reports for the American Petroleum Institute. Papers on the perils of reducing emissions by both Brian Fisher and David Montgomery of Charles River Associates were published in a book by the Competitive Enterprise Institute (Jonathan Adler (ed.), *The costs of Kyoto: climate change policy and its implications*, Competitive Enterprise Institute, Washington, January 1997). When Fisher left ABARE in 2006, he did so to join Charles River Associates.

Embassy deputy chief of mission.[‡] O'Sullivan went on to serve as Prime Minister Howard's international affairs adviser and in 2005 was appointed director-general of ASIO.

The 1997 Countdown to Kyoto conference in Canberra was planned at the November 1996 meeting in Washington, but Evans's links with right-wing anti-environment groups in the United States went back to the early 1990s. Burton records the Competitive Enterprise Institute's R.J. Smith saying that Evans:

> originally heard about us and … set it up with some of our good friends at BHP, Roger Nelson in San Francisco, Nick Allen in Melbourne and others, to bring me down, I think in 1993. I came out for the first time and spent a month at the Tasman Institute [a right-wing think-tank] … and they put me on a tour all around the country to lecture.[30]

In October 2000, Evans spoke at a meeting of the Exxon-funded Cooler Heads Coalition,[31] a project, as we have seen, of the Competitive Enterprise Institute.[32] In 2002, a website of the Cooler Heads Coalition wrote:

> The Lavoisier Group provides the principal intellectual and organizational opposition in Australia to Kyoto and was organized by our colleague, N. Ray Evans of Melbourne. [A critic] accuses the Lavoisier Group of painting the UN's global warming negotiations as 'an elaborate conspiracy in which hundreds of climate scientists have twisted their results to support the "climate change theory" in order to protect their research funding.' Sounds plausible to us.[33]

---

‡ According to the CEI's report on the conference, O'Sullivan 'explained why Australia is fighting against a treaty that imposes an undue burden with minimal return' (<http://www.cei.org/gencon/003,02757.cfm >). See also Burton & Rampton, *Thinking globally, acting vocally*.

Like other denialists, Evans explains away the consensus view of climate scientists by their desire for personal advancement: 'The careers of people like Graeme Pearman have been built on global warming. It's very easy to move step by step into a situation where you are so deeply compromised that you can't wriggle out of it'.[34] This craven behaviour contrasts with the lonely but noble fight of a handful of individuals who, 'like Galileo', hold out against the erroneous view of the masses. Evans is nothing if not indefatigable. As late as May 2006 he was issuing a paper titled *Nine lies about global warming*, which revisited all of the old arguments used by denialists, and in February 2007 a warmed-over version retitled *Nine facts about climate change* was launched by Arvi Parbo in Parliament House. Evans's predilection for adopting entrenched positions is indicated by the view he expressed in 2005 about the outcome of the Vietnam War: 'The only thing wrong with the American campaign was they didn't try and win'.[35]

THE LAVOISIER GROUP

The Lavoisier Group was formed in 1999 ostensibly to bring rationality to a debate dominated by 'green extremism'. It took its name, rather presumptuously, from the founder of modern chemistry, Antoine-Laurent Lavoisier. The group was developed by Hugh Morgan and Ray Evans. Evans serves as its secretary and has organised its conferences. Morgan speaks at its events and WMC had close links. The Lavoisier Group's board includes Ian Webber, a former director of WMC and Santos; industrialist Harold Clough, whose money has supported a range of right-wing organisations (he is also a former director of WMC and current board member of the Institute of Public Affairs); and Bob Foster, a former BHP executive. The Lavoisier Group shares a postal address with the H.R. Nicholls Society.

The president of the Lavoisier Group is the irascible former Labor minister and Western Australian farmer Peter Walsh. Aynsley Kellow, a political scientist from the University of Tasmania, also has connections with the group and spoke at its 2001 conference. It is not known whether, since the Kyoto Protocol came into force, he has changed the title of his 2002 book *The failure of the Kyoto process*. Lavoisier Group conferences feature all of the usual sceptics including Ian Plimer, Bob Foster, Garth Paltridge and Ian Castles. That they should associate themselves with an extreme group like this undermines their credibility as scientific commentators but perhaps reflects their isolation from the mainstream scientific community.†

The group's stance, a strange mixture of conspiracy and apocalypse, was on full display at its inaugural conference in May 2000. Peter Walsh declared that 'the modern CSIRO is not based on science but politics'. Another supporter, Tony Staley, then president of the Liberal Party, described global warming as a form of 'political correctness' and echoed the Bible by saying that they must spread the message for 'the truth shall set us free'.[36] Alan Oxley, a libertarian trade specialist with links to Exxon-funded organisations, told the conference that the Kyoto Protocol was 'a formula for impoverishment', a claim that even the most pessimistic economic modelling backed by the fossil-fuel lobby cannot sustain. One board member of the Lavoisier Group said that ABARE, whose modelling work was funded in large measure by the fossil-fuel lobby, had been captured by environmentalists!

At the 2000 conference Morgan was particularly critical of the Federal Government's Australian Greenhouse Office which, among many other things, had prepared four discussion papers on the pros

---

† In March 2007 Bob Carter, another denialist favoured by the Lavoisier Group, was scheduled to share a platform with Pauline Hanson, prominent Holocaust denier Richard Kreger and well-known neo-Nazi activist Welf Herfurth. See Greg Roberts, 'Hanson's date with Holocaust denier', *The Australian*, 15 February 2007.

and cons of proposals for a system of emissions trading. He compared the papers to Nazi propaganda, describing them as '*Mein Kampf* declarations'.[37] The international negotiations leading to the Kyoto Protocol were seen by the Lavoisier Group as flowing from a conspiracy by climate scientists around the world to sustain their research funding. This has presumably led them systematically to falsify their results, a massive scientific fraud that has escaped the checks of the professional review process because the editors of the prestigious journals are in on it. With evangelical fervour, the Lavoisier Group has conducted a sustained campaign to muddy the waters on climate science and thereby to support the Federal Government's recalcitrant stance.

The Lavoisier Group's submission to a Senate inquiry in 2000 painted a picture of the imminent destruction of Australian sovereignty that would follow from ratification of the Kyoto Protocol, going so far as to compare it to the planned invasion of Australia by Japan: 'With the Kyoto Protocol we face the most serious challenge to our sovereignty since the Japanese Fleet entered the Coral Sea on 3 May, 1942'.[38] In words that could have been penned by an ideologist of the Montana Militia, the submission talked of the Kyoto Protocol ushering in a 'new imperial order', of the 'termination' of our sovereignty as a nation, of 'imperialists ... in green clothing', and of 'the threat of invasion'. It conjured fears of the 'police powers' of an unaccountable Kyoto secretariat based in Bonn, 'an international tribunal which ... will have the power to transfer, or destroy, wealth and income within Australia on a massive scale'. It claimed that 'our sovereignty will be relocated from Canberra to Bonn', and feared that the World Trade Organization, subjugated and corrupted by the demands of the Kyoto Protocol, would become 'an instrument of imperial authority'.

The submission suggested seriously that: 'Australia will only be able to escape from entrapment in this new imperialism through immense political upheaval of the kind experienced by George Washington and his colleagues when they rebelled against the authority of the British Crown and established the United States'. Such is the tone of conspiracy and apocalypse in the Lavoisier submission that one could be forgiven for mistaking Lavoisier for LaRouche.

More generally, one can find the following arguments in the various papers promoted by the Lavoisier Group:

- There is no evidence of global warming.
- If there is evidence of global warming, then warming is not due to human activity.
- If global warming is occurring and it is due to human activity, then it is not going to be damaging.
- If global warming is occurring, it is due to human activity and it is going to be damaging, then the costs of avoiding it will be too high, so we should do nothing.

It is impossible to have a rational discussion with such people, for they are immune to evidence and argument.

Like Hugh Morgan, the Lavoisier Group has a special antipathy for the Australian Greenhouse Office, which it sees as staffed by a power-hungry bureaucracy who would sell out Australia's sovereignty to their counterparts in Europe.[39] In May 2003 the group wrote an open letter to the Prime Minister reminding him of his anti-Kyoto pledge and alerting him to the fact that the AGO 'and their colleagues in other departments' (which would have included Treasury) were planning to conduct research into the consequences for Australia of the introduction of a carbon tax or an emissions trading scheme. Signed by the president, Peter Walsh, the letter concluded: 'We urge you to back your judgement and your concerns with a directive to the bureaucrats who seek to undermine your policy to cease and desist'. For an organisation that claims to have a case built on facts, including the economic effects of reducing emissions, it is remarkable that it should intervene so strongly to prevent others from gathering them.

It is tempting to dismiss the Lavoisier Group as a bunch of crackpots, but its claims have undoubtedly influenced some Coalition ministers and back-benchers, among whom a strong anti-European sentiment has developed. According to one consultant and industry adviser on greenhouse gases: 'I think the sceptics have had an impact. I think Australia's reluctance to ratify the Kyoto Protocol has come down to the tactics of these groups that are supported by industry'.[40] The Liberal Party chair of the parliamentary committee to which the Lavoisier Group made its submission, Andrew Thomson, seems to have been swayed by the arguments. On 17 April 2001, he wrote a letter to *The Australian Financial Review* defending the Lavoisier Group and Peter Walsh.[†]

What is most striking about the Lavoisier Group is its lack of intellectual sophistication. An independent observer at the group's 2001 conference summed up the audience as follows:

> The typical Lavoisier Group supporter is over 60 and male, lives in Melbourne, was a scientist or engineer who worked for a large mining company, and has conservative views. The ABC is 'in the enemy camp', announcements that a politically incorrect view is about to be put are greeted with guffaws, and to a man the human race is 'man'.[41]

The only way an organisation with such crude opinions could have political influence is for the politicians it targets to share the

---

[†] Thomson lost his party endorsement for the blue-ribbon seat of Wentworth soon after.

same loathing of anything environmental, the same emotional affiliation with mining and the same apocalyptic vision of the consequences that must accompany reductions in greenhouse gases.

BUREAU OF METEOROLOGY

One of the more curious aspects of the climate change debate in Australia has been the almost complete absence from it of the Bureau of Meteorology. Until recently the bureau has been largely silent or, when compelled to comment, has cast doubt on the link between changes in climate and the enhanced greenhouse effect. The contrast with its counterpart at the CSIRO, the Division of Atmospheric Research, has been striking. More recently there has been some shift in this position under its new director, Dr Geoff Love, who is quoted on the bureau's website as follows: 'I expect climate change to affect all Australians. It is the Bureau's responsibility to provide decision makers and the general public with accurate observations and information about our changing climate'.[42]

Few would disagree, which raises the question of why the bureau failed in its responsibility for so long. The answer lies not in political interference but in the influence of the head of the organisation, Dr John Zillman, who served as director of meteorology for 25 years until retiring in 2003.

Zillman is an eminent meteorologist who for several years was the vice-president then president of the World Meteorological Organization, which was heavily involved in the formation of the IPCC. Despite his very active involvement in the work of the IPCC, Zillman has been described by a senior Australian climate scientist as a 'closet sceptic'.[43] In April 2003 he told a House of Representatives committee that he rejected the view that the severity of the drought was due to climate change.[44] Sitting next to him, his deputy, Dr Mike Manton, told the committee that the observed warming trend could not be attributed to 'greenhouse climate change'. He also asserted that nothing could be said about the relationship between climate change and regional variations in climate. Zillman then said that the Murray-Darling Basin had been *cooling*, with a net decline in temperature of half a degree.

These claims were strongly challenged by Dr David Karoly, an eminent Australian climate scientist and professor of meteorology at the University of Oklahoma, who pointed to a range of published studies that contradicted them.[45] He argued that the claim that the Murray-Darling Basin was cooling was not only incorrect but was also contradicted by the bureau's own data, with the mean temperature showing a warming.

Two months earlier, Julie Macken had written a story in *The Australian Financial Review* quoting Karoly on the politicisation of the bureau: 'They [the bureau] are concerned, politically, that science doesn't make this connection, primarily because the bureau does not want to alienate its political masters in Canberra'.[46] Yet it is more likely that the Government did not need to exert political influence over the bureau's scientific research because Zillman was doing it unprompted. Former senior scientists at the bureau have confirmed that they engaged in 'self-censorship' because Zillman had been a sceptic and, even after he had accepted the likelihood of human-induced climate change, continued to stress the uncertainties and the need for caution in all claims.

In 2004 Zillman accepted an invitation from Hugh Morgan to launch a book by denialist William Kininmonth titled *Climate change: a natural hazard*. Kininmonth refers to the belief in human-induced global warming as 'self-delusion on a grand scale'. The Lavoisier Group was the organiser of the launch.

Kininmonth's book was a sustained attack on the credibility of the IPCC and, since Zillman had a long-standing and active involvement in the panel, he was bound to defend its reputation, which he did. But in agreeing to launch the book, Zillman endorsed the view that climate change denialists are an important and valid part of the scientific debate, something that the fossil-fuel lobby had been actively promoting.[47]

Speaking at the launch, Kininmonth expressed his sincere thanks to Ray Evans for his 'assistance and encouragement'. It is odd, to say the least, that Zillman should associate himself with the Lavoisier Group, Hugh Morgan, Ray Evans and Bill Kininmonth, those at the very centre of climate change denial in Australia and who have actively worked to confuse the public about climate change science.

11. THE BATTLE FOR PUBLIC OPINION

INVASION OF THE SCEPTICS

By the time global warming loomed as the most alarming threat to humanity, environmentalism had given rise to its opposite, a virulently hostile coalition of industrialists, right-wing commentators and conservative politicians. This led to some unlikely alliances, none more bizarre than that revealed in July 1998 when the ABC screened a two-part UK documentary about environmentalism entitled *Against nature*. According to the publicity material, the documentary characterised 'environmentalist ideology as unscientific, irrational and anti-humanist'. *Against nature* had created a furore after being broadcast on Britain's Channel 4 in December 1997, not least for its extraordinary claims that modern environmentalism had its roots in Nazi Germany and that self-interested environmentalists were responsible for enormous suffering in the Third World.

Of course, there is nothing wrong with a documentary that challenges accepted views about the need to protect the environment, including views about global warming, but *Against nature* was a series of distortions rather than a sceptical commentary. Employing a range of editorial devices designed to exclude debate and inflame passions, the program fell well short of accepted journalistic standards. Images of Third World children dying of horrible illnesses were accompanied by commentary about environmentalist opposition to dams bringing clean water and electricity, a contrast which seemed designed to portray greenies as callous fanatics. Environmentalists were blamed for preventing construction of the Narmada Dam, although the dam was actually stopped by the Indian Supreme Court after a protest campaign by local residents who would have been displaced by it.

The producers claimed that the program 'highlight[ed] the absence of scientific rigour behind notions like the greenhouse effect and global warming'. The views of environmental activists were contrasted with those of scientists who did not believe in the enhanced greenhouse effect, thereby characterising the global warming debate as 'opinion' on one side and 'fact' on the other. And while the affiliations of the environmentalists were shown, the anti-environmentalists were presented as independent experts. One of the most authoritative voices, Fred Singer, was a well-known greenhouse sceptic who today heads an organisation supported by oil companies.[1] He is also an adjunct fellow at the Frontiers of Freedom Institute.[2] Other US interviewees were closely associated with the Wise Use movement, whose founder, Ron Arnold, has declared that his objective is to destroy the environmental movement, which he regards as aligned with communism.[3] It was as if the ABC were to screen a program reporting a doctor's claim that smoking does no harm, yet fail to mention that he had been funded by the tobacco industry.

After the UK screening, four of the environmentalists interviewed on the program complained to the Independent Television Commission. The commission ruled that 'the editing of the interviews had ... distorted or misrepresented [the] known views' of the environmentalists and that 'the production company had misled them ... as to the format, subject matter and purpose of the programmes'. Channel 4 subsequently made a lengthy on-air apology and put together a discussion program that gave greens an opportunity to respond.

The screening of *Against nature* here and in the UK was a considerable coup for right-wing groups that view environmentalism as a threat to capitalism and freedom. But the most remarkable feature of the documentary emerged only after it was shown in Britain. Detective work by *Guardian* columnist George Monbiot revealed that the program's makers were linked to an obscure political group named the Revolutionary Communist Party (RCP).[4] The RCP published a controversialist journal entitled *Living Marxism* (later renamed *LM Magazine*), which frequently ran bitter attacks on environmentalism, describing it as a middle-class indulgence or a neo-colonial smokescreen. Its articles had titles like 'Red and green won't go', 'Animals have no rights' and 'Environmental imperialism'. The journal also took controversial positions on international issues, including opposition to sanctions against the apartheid regime and to the ban on landmines, and support for the Bosnian Serb forces and the Hutu militias.

Against nature was a kind of outgrowth of *LM Magazine*. The key interviewee of the documentary was Frank Furedi (operating under the pseudonym Frank Richards), the founder and president of the RCP. Furedi, now a professor of sociology at the University of Kent, wrote regularly for the magazine, as did another interviewee, John Gillott, while a third, Robert Plomin, was interviewed sympathetically in its pages. The producer and director of the documentary, Martin Durkin, also had close links with the party (although he denied any such connection).

None of these revelations has stopped the rise and rise of Frank Furedi, the intellectual force behind *Against nature* and the RCP. Since then he has written a number of books that explore and denounce the excessive emphasis in Western culture on risk and danger, the most recent being *Politics of fear: beyond left and right*.[5] Some of his arguments are plausible and valuable, yet there is a deeper purpose to them. The blanket criticism of over-sensitivity to risk is applied specifically to climate change, as if global warming were just another invention of 'doom-mongers'.[6] In visits to Australia, this has made Furedi an attractive guest for right-wing broadcasters such as Michael Duffy, who has twice featured him on his ABC radio program *Counterpoint*,[7] the Centre for Independent Studies, and more neutral venues such as the Brisbane Writers Festival. Despite being courted by the right, the former Trotskyist claims that his politics have not changed.[8] The successor to *LM Magazine* lives on in cyberspace in the form of the e-zine *Spiked Online*, a peculiar mix of ultra-libertarianism and 'left-wing' opinion. It often carries articles by Furedi.

Lumping global warming in with a number of other social anxieties in order to discredit them in toto is still a preferred tactic. Such an approach was used by *The Australian* newspaper as late as October 2006 to undermine the credibility of the Al Gore documentary *An inconvenient truth*. Categorising it as science fiction, the Murdoch broadsheet editorialised: 'the future is not going to be anything like the dystopian worlds depicted in Fritz Lang's *Metropolis*, Stanley Kubrick's *A clockwork orange* or Al Gore's *An inconvenient truth*'.[9]

And in February 2007, Peter Saunders from the Centre for Independent Studies wrote an opinion piece ridiculing a study linking consumption of tofu to cancer and referring to Furedi's argument that we have become soft and fearful. He then listed global warming among other 'media panics', such as bird flu and binge-drinking.[10]

BJORN LOMBORG

The Danish statistician Bjorn Lomborg represents the more respectable end of the denialist-sceptic camp. His 2001 book *The skeptical environmentalist* was a sustained and apparently carefully documented attack on environmentalism.[11] This previously obscure academic clearly struck a powerful nerve with his claim that environmentalists have grossly exaggerated the state of environmental damage and that, in truth, things have never been better. His book set out to use the statistical evidence to demolish what he called the 'litany' of four big environmental fears – exhaustion of natural resources, overpopulation, extinction of species and worsening pollution.

The media on both sides of the Atlantic lapped it up. Major features appeared in *The Economist*, *The Daily Telegraph*, *The Washington Post*, *Time* magazine and *Business Week* among dozens of others. *The Daily Telegraph* in London called *The skeptical environmentalist* 'probably the most important book on the environment ever written' and the liberal *Washington Post* lauded it as 'the most significant work on the environment since the appearance of its polar opposite, Rachel Carson's *Silent Spring*, in 1962'. Even *The Sydney Morning Herald* carried a sympathetic story.

Long-time observers recognised *The skeptical environmentalist* as a recycled version of the arguments and 'evidence' put forward by environmental sceptics such as Julian Simon and Gregg Easterbrook, both of whom had produced better written and more persuasive tomes. (It was not until May 2006 that Easterbrook felt able to write: 'based on the data I'm now switching sides regarding global warming, from skeptic to convert'.[12]) Lomborg's book was a useful reminder that the environment movement around the world has won some major victories, not least in forcing governments in rich countries to sharply reduce air and water pollution. But the extraordinary publicity given to *The skeptical environmentalist* revealed something important about the modern politics of the environment: a small but influential minority detests environmentalists. The book provided an opening for the hidden seam of anti-environmentalism to surface, and it did so with a vengeance.

The motivations of the anti-greens are not always clear, but we sometimes see them emerge after bushfires or land-clearing restrictions, when right-wing columnists and letter-writers launch splenetic attacks on greenies as somehow being responsible. For some, environmentalists are seen as no more than self-righteous moralisers who play on guilt to win political influence. Others see environmentalism as the enemy of capitalism because it often points to the damage that corporations and the market can do if left unchecked.

The skeptical environmentalist played particularly well in the United States, where the far right regards greenies as the nearest thing to Satan, but Lomborg has been much sought after by anti-greenhouse groups in Australia too.[13] In October 2003 he arrived for a speaking tour sponsored by the Institute of Public Affairs but apparently initiated by Ian Macfarlane, the federal industry minister and climate change sceptic, whose spokesperson said Lomborg 'spoke a lot of sense'.[14] He visited again in August 2006 to promote his next book, *How to spend $50 billion to make the world a better place,* which ranked global warming well down the list of world problems and provided him with an opportunity to talk down the threat. He is said to have visited 'courtesy' of the Centre for Independent Studies.[15]

In Australia, right-wing commentators seized on *The skeptical environmentalist*. In an outburst against the 'loony greens', Miranda Devine eulogised Lomborg for demolishing 'the myths which form the basis of the green pseudo-religion' and declared the debate on Channel 9's *60 Minutes* between Lomborg and Peter Garrett as a 'turning point in Australia for logic over emotion'. Lomborg, she wrote, is the best example of a 'New Enlightenment' which relies on empirical evidence and established facts to prove that there is 'an absolute truth'. Writing in *The Australian*, Alan Wood declared that Lomborg had shown up environmentalists, teachers and the media as 'no more than the dupes of political and ideological causes dressed up as impartial science'.[16]

Lomborg's success drew fire from environmentalists and scientists, with forceful critiques being published in prestigious journals such as *Scientific American* and *Nature*. Perhaps the most damaging blow came from an official Danish committee that accused Lomborg of 'scientific dishonesty'.[17] As the criticism by some of the world's leading scientists gathered momentum, Lomborg increasingly sought support from the far right. This was a tactical mistake for someone who claimed to be a dispassionate analyst. While Lomborg seemed to be a godsend for the right, the alliance was a decidedly odd one. After all, Lomborg was a young, gay, vegetarian Dane who said he continued to vote socialist and support a strong welfare state. He also claimed to have been a Greenpeace activist, although the organisation said it had no record of his membership.

Lomborg's claims about climate change were among the most contestable in his book. Like other sceptics, he focused his attention on the uncertainties that surround climate science and accepted uncritically the more apocalyptic claims about the economic costs of cutting emissions. Consequently, Lomborg himself was guilty of gross overstatement of the end-of-the-world variety. He wrote that trying to limit global warming 'would require a complete cessation of all carbon emissions by 2035, essentially shutting down the world as we know it'. Lomborg's central argument is this: 'Most forms of environmental pollution either appear to have been exaggerated, or are transient – associated with the early phases of industrialisation and therefore best cured not by restricting economic growth, but by accelerating it'.[18]

It is true that many forms of pollution have been cleaned up in industrialised countries, although there are several crucial exceptions to this, with high-level nuclear waste, plastic wastes and greenhouse gas emissions perhaps the most obvious. Factories can no longer dump toxic wastes into the rivers willy-nilly (although some still do and their owners find themselves in court), and air pollution in most major cities is not as bad as it was two or three decades ago. These laws have been forced on industrialists, usually in the face of denial, obfuscation and outright opposition. It was not the car industry that initiated the push for lead-free petrol; it was the government at the behest of the citizenry. Thus while Lomborg concludes that improvements to the natural environment render environmental activism unnecessary, in fact the gains from such activism prove its effectiveness.

A nation's political system, rather than its wealth, determines its response to environmental threats. By noting the association between improvements in some forms of pollution and high incomes, Lomborg confused correlation with causation, an unforgivable error for a statistician. Improvements in the environment do not follow naturally from more economic growth. Keeping pollution and toxic wastes relatively low and reducing them further requires an unceasing battle against the effects of economic growth. Laws requiring cars to have catalytic converters made a big difference, but most experts believe that the sheer growth in the volume of cars is now starting to offset the benefits of better technology, and urban air pollution is expected to rise again. Things will become worse unless environmentalists and the citizenry in general insist on even tougher standards. Change will not just happen automatically as incomes rise, not least because of vigorous opposition to regulation from corporate interests.

In Australia, some oil companies have strenuously resisted tougher fuel and engine standards planned by the Federal Government, and in the United States the Bush Administration wound back some of the environmental improvements introduced by the Clinton Administration, despite a healthy economy. And even the small step on the path to reducing greenhouse gases represented by the Kyoto Protocol has been fought tooth and nail by powerful vested interests. The governments of the United States and Australia have done all they can to sabotage any agreement. Perhaps Bjorn Lomborg would like to nominate the level of national income at which he expects Australia and the United States to abandon their opposition to the Kyoto Protocol.

GLOBAL SPIN

The denialists have been diligent in bringing a range of international figures to Australia to bolster their case. Mark Steyn is a witty right-wing Canadian newspaper commentator whose columns appear in *The Spectator*, *National Review* and, in this country, *The Australian* among others.[19] During his visit in August 2006 most of his commentary concerned Western civilisation, Islam and so on, but he is also an ardent climate change denier. On Michael Duffy's ABC program *Counterpoint*, he said he found worry about climate change 'very curious'. He argued that sea-level rise will not sink the Maldives for hundreds of years and that when that happens we can move the population to the south of France, 'and being Sunni Muslims they'll fit right in by the year 2500 because everyone else in the south of France will be Sunni Muslims'. Worrying about climate change 'is a sign of decadence'.[20]

To a rapt audience of conservatives in Melbourne, Steyn delivered the Institute of Public Affairs C.D. Kemp Lecture. Winning the quinella of right-wing think-tanks, he went on to speak at the 'big ideas forum' of the Centre for Independent Studies.[21] Those attending his speeches included Government ministers Alexander Downer, Nick Minchin and Santo Santoro.[22] It emerged two months later that Steyn's visit to Australia was funded by the taxpayer. The Department of Foreign Affairs and Trade paid $12,023 to bring him to Australia under a program aimed to persuade foreign journalists to write more favourably about Australia, although Steyn is not a noted critic of the Howard Government.[23]

Among the more colourful characters brought to Australia to bolster the denialist case was President Putin's economic adviser Andrei Illarionov. His sponsor, the Institute of Public Affairs, was well aware that Illarionov had been campaigning vigorously against Kyoto for some years. Like those of the Lavoisier Group, Illarionov's pronouncements had an apocalyptic tone. A talk given shortly before his Australian visit in 2004 was entitled *The Kyoto Protocol: an assault on economic growth, environment, public safety, science and human civilization itself*. Illarionov did not blame Kyoto for obesity, paedophilia and Islamic fundamentalism, but he did write in *The Moscow Times* that it was killing off the world economy like an 'international Auschwitz'. In the London *Financial Times* he compared it to fascism and communism because it was 'an attack on basic human freedoms behind a smokescreen of propaganda'. The aim of this 'death pact', which his President and the Duma had just endorsed, 'is to strangle economic growth and economic activity'. Naturally, *The Australian* turned over its opinion page to the visiting Russian, where he fulminated against 'fraudulent science', 'pseudo-scientific fabrications', the 'nonsense of global warming', 'a supranational bureaucratic monster', and Kyoto as 'a sort of international Gosplan, a system to rival the former Soviet Union's'. For good measure, in case anyone had missed the point, he declared 'Kyoto-ism' to be 'one of the most aggressive, intrusive, destructive ideologies since the collapse of communism and fascism'.[24]

As we have seen, astroturfing is a well-established industry campaign method in the United States, and has been practised with vigour by opponents of the Kyoto Protocol, with much of it traceable to organisations funded by ExxonMobil.[25] The technique has been used in Australia only sporadically. However, in April 2005 a half-day workshop in Melbourne featured Canadian PR consultant Ross Irvine, dubbed the 'anti-activist activist'. The workshop, titled 'Activists: how to beat them at their own game', was sponsored by the Institute of Public Affairs and the Public Relations Institute of Australia (PRIA). The involvement of PRIA was curious because astroturfing is condemned by the more respectable end of the PR industry. Kath Wilson, a journalist, paid the $599 fee to hear what Irvine had to say.[26] Wilson reported that the attendees were a who's who of industry and government PR people, including representatives from Rio Tinto, Shell and Dow Chemical.

Wilson recorded Irvine describing corporate responsibility as 'weakness'. 'Quite frankly, business doesn't have the resources and capability that activists do', he said, before urging participants to set up citizens' front groups. David Hawkins of PRIA chimed in with: 'The challenge, I think, from what Ross is saying, is ... we need to work out how we *can* break the law to do these things'. At least two of the participants in the Melbourne workshop absorbing Irvine's wisdom were Federal Government staffers, including David Gazard, who works for Treasurer Peter Costello.

In the United States, the denialist camp recruited popular author Michael Crichton, whose novel *State of fear* characterised global warming as the fabrication of nefarious environmentalists. Risibly, in Australia Michael Duffy used Crichton's novel as an authority for his own denialist position. 'Crichton believes green groups have invented this crisis to attract members and money. For the greens, no crisis means no cash.' Adopting the denialists' standard terminology to dismiss thousands of articles in peer-reviewed scientific journals, Duffy wrote: 'The book is a fascinating exposé of one of the great scams of our time. And, unusually for a work of fiction, *State of Fear* has footnotes and 22 pages of references to support its claim that global warming is politicised junk science'. Footnotes can turn fiction into fact, it seems. Perhaps Duffy's grandchildren will take comfort from his final words: 'So next time there's a sweltering day, think about *State of Fear* and just lie back and enjoy the heat'.[27]

Most of the claims made by commentators such as Michael Duffy, Alan Wood and Andrew Bolt can be traced to literature emanating from the United States, and particularly from organisations that have received funding from ExxonMobil. Not only are the 'authorities' the same – Fred Singer, Willie Soon, Richard Lindzen – but so too is the language in which their arguments are presented. Borrowing from the tobacco lobby, these writers contrast the 'junk science' of the IPCC with the 'sound science' of the denialists. They portray the environment movement as a green religion and a multi-million-dollar 'panic industry'. The scientific community is characterised as a closed, secretive clique that condemns scepticism and is hostile to criticism. They talk about imprecise modelling and the problems of prediction, and prefer anecdotes such as repeated reference to the 'little ice age'. They reinforce the Australian Government's arguments about the futility of acting unless China and India do. An earnest, and typically batty, summary of the claims, the language and the sources in Australia is the 2006 report by Ray Evans for the Lavoisier Group mentioned previously. *Nine lies about global warming* reads like a song sheet for deniers writing opinion pieces, one that Miranda Devine had propped in front of her as she typed out an article for *The Sydney Morning Herald* in March 2006 in order to dismiss 'hysterical claims' that sea levels are rising and to attack the 'near-religious fervour behind much of what passes for debate about climate change'.[28]

PUBLIC OPINION

At the time of the Kyoto agreement in 1997, a large majority of Australians supported measures to cut emissions and wanted their country to play a positive role in international efforts. In the 1980s and 1990s Australia had developed a reputation as a leader in sustainable development. This was a legacy in large measure of the Hawke

Government, spurred on by an active and effective environment movement. The momentum was lost when Paul Keating became prime minister in 1991; his Government showed little commitment to climate change. The Howard Government, elected in 1996, wanted a radical break from the Labor era and environmentalism was seen to be Labor territory. The outcome of the Kyoto conference at the end of 1997, while celebrated as a great victory by the Government, left most Australians feeling uneasy. After all, world environmental leaders were not supposed to rejoice when they had, through threats, been let off the hook and avoided their environmental responsibilities. The cartoonists captured it best.

The unease turned to serious discomfort as the Howard Government began to signal that even the best deal in the world was not good enough. The feeling was reinforced by a report by the Australia Institute in 1999 which concluded that Australians have the highest level of greenhouse gas emissions per person of any industrialised country, a story that appeared on the front page of *The Age*.[29] It was a simple but powerful fact that instantly took root and was quoted endlessly in Australia and abroad.

But it was one thing for Australians to express concern about climate change and another to do anything about it, and it took nine years before their anxiety ran deep enough for them to think about changing their votes. For that period most Australians were in a state of denial about the implications of global warming, simultaneously knowing and not knowing. In report after report, scientists rang the alarm bells ever more loudly, yet Australians would not connect their political views or personal behaviour to the emerging crisis they were hearing about. The problem seemed so huge and so intractable that it did not bear thinking about. The scientists could not agree, it seemed, so let's not worry until we really have to. Besides, the effects are a long way off and the answer will be found before it is too late.

For much of the global warming campaign, environmental organisations in Australia have been ineffective. It was not until some years after the Kyoto conference that the Australian Conservation Foundation, preoccupied with forests and biodiversity, overcame organisational inertia and placed climate change at the top of its campaign agenda. Like all organisations it was slow to adapt; its staff, council and membership had a great deal invested in its historical campaigns and often had entrenched positions.

Greenpeace was the most advanced of Australian groups, principally because the international organisation was leading the campaign on climate change in Europe, where the debate has been years ahead of Australia. But even here, Greenpeace seemed to believe that Australia was not important in the global debate and was slower than it should have been to shift away from traditional concerns about uranium mining and whales and to recognise the centrality of climate change. Its first major campaign on the question was against a proposed oil shale development in Queensland, which was a side issue.

The third major organisation, WWF-Australia, established a climate change unit only in 2002, when it was apparent that the Australian arm of the organisation lagged well behind its international affiliates.

In their campaigning, Australian green groups have been locked into an Enlightenment view of human psychology – if only people had more information, they would understand that they must act. But the problem was not one of insufficient knowledge; it was one of connecting the knowledge with people's sense of responsibility to act in their personal lives, and, even more importantly, to act politically. The Government skilfully shifted responsibility from itself onto the consumer with a series of campaigns and policies to promote voluntary schemes. Drive less, it said. Change your light bulbs; think about buying green power. The real message was, 'It's not our fault, it's yours, so don't change your vote'. The public creased its brow over newspaper stories about the effects of warming, then turned the page and became outraged at rising petrol prices.

A Newspoll survey in April 2001 found that 80 per cent of Australian adults supported ratification of the Kyoto Protocol without the United States if necessary.[30] However, it would be another five years before public concern about climate change sank in deeply enough for it to become a major political issue.

MEDIA SLUMBER

One important cause of this disconnection was the media's long-term failure to give climate change the attention it deserved. In part this was due to ignorance, and in part to the Government's successful spinning. The situation was not helped by the practice of rotating young journalists out of the environment beat just as they began to understand the issues. Rookie reporters had no benchmarks or historical understanding against which to judge the claims coming from the Government, and it got away with outrageous claims as a result. Few if any of the reporters undertook the task of examining Australia's annual greenhouse gas inventory, which is easy to understand and paints a picture of manifest policy failure every time it appears. We have already seen, too, how journalists accompanying the Australian delegation to international climate change conferences often seemed to take the view that it was their patriotic obligation to report the Government's line uncritically, no matter how fanciful the arguments advanced.

Environment stories are generally low on the list of newsworthiness and only come into their own as novelty or travel stories. Climate change was seen as boring and this made it nigh-impossible to persuade the media of the overwhelming implications for the future of the Australian economy and way of life. The lack of media interest and understanding of the issue – at least in Australia – was powerfully illustrated in 2004 when London's *Observer* newspaper broke the story of a leaked report commissioned by the Pentagon on the security implications of climate change.[31] More alarming than any claim made hitherto by environmental activists, the report argued that Australia and the United States 'are likely to build defensive fortresses around their countries' to protect their resources from desperate outsiders and aggressive states created by rapid and unpredictable climate change. It analysed the prospects for aggression 'if carrying capacities everywhere were suddenly lowered drastically by abrupt climate change. Humanity would revert to the norm of constant battles for diminishing resources ... Once again, warfare would define human life'. The Pentagon report was commissioned by a senior Pentagon adviser, Andrew Marshall, who in his decades in the defence department has acquired the status of a guru. It marked a dramatic shift in the international debate over climate change, with defence and intelligence experts at the highest level becoming increasingly alarmed at the implications for global security. Yet the Australian press showed no interest at all despite Australia's prominent place in the Pentagon analysis. By any editorial standard it was a huge story; yet somehow editors in this country passed over it, which must have elicited a huge sigh of relief in the Prime Minister's office.

More generally, the media were reluctant to link weather events to climate change. Acres of newsprint were devoted to the El Niño drought of 2002, the continuing water crisis and the horrendous bushfires of 2002–03, with virtually no mention made of climate change, even though Australia had some of the world's leading climate scientists who had published papers on the topic and who would have been willing to talk about the links if only a journalist were to ring.

There were some notable exceptions to this studied ignorance. A handful of journalists had been reporting the issue for years, sometimes in the face of indifference or hostility from their editors. Claire Miller of *The Age*, Julie Macken of *The Australian Financial Review*, Sarah Clark at the ABC and, in more recent years, Wendy Frew of *The Sydney Morning Herald* deserve recognition. Mention should also be made of Murray Griffin, the editor of the independent newsletter *Environmental Manager*, which has followed the issue closely for years, and Alan Tate of the ABC and Gavin Gilchrist of *The Sydney Morning Herald*, who pursued the issue vigorously in the 1990s.

But commercial television and the tabloids virtually ignored the topic, and even in the broadsheets and on the ABC, climate change was treated as just another green issue, with rabid environmentalists pitched against self-interested businesses. The exception was *The Australian*, which for years had been conducting a sustained war on climate science and the Kyoto Protocol. As the journalistic home of right-wing ideas, *The Australian* has campaigned vigorously against the influence of 'post-modernism' throughout Australia's institutions. It has railed against the 'black-armband' interpretation of Australian history, multiculturalism, post-modernism in school curricula and the moral relativism of the left. It sees itself as standing for objective truth after decades of leftist challenge to the Western canon. Yet in the case of climate change it has actively promoted those who challenge the established science. The sceptics who inhabit its pages not only dismiss the science but also constantly attempt to 'deconstruct' the motives of the scientists who carry it out.[32] They are always on the lookout for biases and prejudices that could lie behind the scientific facts on global warming and turn them against those who want to act on the evidence. In their view, scientific truth is malleable, contingent and contestable. Like the creationists who believe that victory requires them to destroy the theory of evolution, *The Australian* promotes a form of anti-scientific fundamentalism that has less regard for scientific method than the most committed constructivist on any university campus.

Modernism now finds itself under siege from both the dwindling band of academic post-modernists and resurgent neo-conservatives. Both reject the claims of science to objective truth. For the former the truth of modernism was socially constructed and the real truth is always contestable; the latter never accepted the elevation of matters of fact over matters of belief. For the sceptics and their patrons at *The Australian*, loyalty to belief is paramount and every piece of evidence that challenges their convictions represents a profound threat to their world view.

After years in a media ghetto, in late 2006 climate change suddenly became a mainstream issue and the drought of reporting and analysis turned into a flood. Although a thoroughly welcome development, some peculiar treatments of the issue began to appear. Soft-porn men's magazine *FHM* managed to secure an interview with Al Gore on his Australian visit, opening the interview with: 'First things first, Mr Gore: could you beat George W. Bush in a fist fight?'[33] *FHM* drove home the climate change message with a companion page picturing two naked blondes below the headline: 'Global warming! As global temperatures rise, expect more snow angels like these to emerge from Sweden's diminishing permafrost'. The *Sun-Herald*'s travel section carried a cover story identifying the ten most spectacular natural sights that may soon disappear as a result of global warming and urging readers to see them while they could.[34] They included the Barrier Reef and Tropical North Queensland. *The Australian*'s glossy *Wish Magazine* followed up in October with a story headed: 'Going, going ... go now. Seven wonders of our world, unique in their natural or cultural splendour, are under threat. See them now, before it's too late'.[35] It provided the dates by which they should be seen, adding to the disturbing sense of fatalism that stories like this invoke. Whereas others see climate change as a serious threat to tourism, the industry has embraced it as a marketing opportunity.

12. COMICAL IAN

BALD HILLS

When he was appointed to the post of environment minister in July 2004, one quality Senator Ian Campbell did not lack was enthusiasm. Some observers were impressed that Campbell, soon after taking on the job, appeared to accept the science of climate change while taking a tour of the ancient forests of Tasmania. He was ten years after the rest of the world, but in the bleak political landscape it appeared as a ray of light. Unhappily, it soon became clear that he accepted the science only when it was convenient to do so. In early 2006, when asked about rising sea levels in the Pacific, he echoed the Prime Minister's words of 1997, claiming that, 'The jury is really out'.[1] This suggested a courageous willingness to select from the compendious analysis set out by IPCC those parts with which he could agree and to reject those which the world's climate scientists had, on balance, got wrong. We subsequently learned that the Government had banned climate scientists from discussing or even mentioning the subject of sea-level rise and environmental refugees. For all of his bumbling we owe a debt to Campbell because his ineptitude has given us a much clearer insight into the Government's real greenhouse agenda than we gleaned from his cleverer predecessors.

Campbell's handling of the Bald Hills wind farm fiasco threw into sharp relief the influence on the Howard Government of the fossil-fuel industries as compared with their renewable energy competitors. In response to the MRET scheme, investment in wind energy had boomed briefly. A company named Wind Power Pty Ltd proposed to construct a large wind farm in South Gippsland in Victoria. Days before the federal election of October 2004, aware of significant local opposition to the proposed development, Campbell issued a media release indicating that, if the Government were re-elected, he would take a dim view of the wind farm proposal.[2] The site was in a marginal seat held by Labor, one the Liberal candidate subsequently won.

Eighteen months later, in announcing his decision to veto the proposal, Campbell claimed that it would pose a risk to a threatened species, the orange-bellied parrot. He said that the proposed wind farm would 'hasten the extinction of that species'.[3] The Bald Hills wind farm had passed all Victorian planning approvals. Based on its studies, the Victorian government held no fears for the orange-bellied parrot arising from the development. Indeed, the parrot had never been sighted within ten kilometres of Bald Hills. The Victorian government's analysis concluded that the best estimate of the expected impact of the proposed wind farm would be one dead parrot every 1000 years.[4]

The minister used his powers under the *Environmental Protection and Biodiversity Conservation Act*, the transformed environmental laws passed with Democrats' support in 1999. The bill was opposed by most of the major environment groups, but supported by WWF-Australia, Humane Society International and the Tasmanian Conservation Trust, conservative groups that have since been favoured by the Government. It is now clear that the main environment groups were right to be sceptical about this Act. The decision to block Bald Hills was only the fourth time in six years that the Government had used the legislation to block a development, despite it being applicable to thousands of proposals, and this time it was used to block a development that would reduce Australia's greenhouse gas emissions.[5]

While Campbell used the Act to stop Bald Hills, he subsequently said that community opposition to the wind farm had been important and that communities should have a say, though few local objectors had ever heard of the orange-bellied parrot. It was later revealed that the minister had ignored his department's advice that there were no significant environmental obstacles to the Bald Hills proposal, although he blocked access to most files sought under freedom of information legislation, claiming that releasing the documents might 'promote ill-informed speculation about the Government's decisions'. It is truer to say that the documents confirmed well-informed speculation about the decision.

Businesses were spooked by the arbitrary application of environmental planning rules, with even the Business Council of Australia weighing in with criticism of Campbell for creating investor uncertainty. Campbell's behaviour was described by *The Australian* as 'bizarre' and the paper editorialised in favour of his dismissal.[6] The proponent of Bald Hills, Wind Power, took legal action against the Government, and, searching for a way out of the hole he had dug for himself, Campbell announced he would consider a revised proposal. This undertook to shift six of the 52 turbines 150 metres so that they would not intrude into a zone two kilometres from the coast, a trivial move that served only as a face-saver for Campbell. The minister approved the revised proposal in December 2006, but Wind Power still planned to sue the Commonwealth to recoup the extra costs caused by Campbell's actions.[7]

The hostility of the Government to the wind industry had already been revealed in the leaked minutes from the 2004 LETAG meeting; but there is a difference between failing to encourage an industry and actively working to thwart its development. Following that meeting, the Government announced that it would not extend the MRET renewables investment scheme. With surprising speed, the 9500 GWh of new renewables by 2010 allowed for under the scheme had been taken up largely by new wind farms, an established technology ready to exploit Australia's abundant wind resource. Soon after the Bald Hills fiasco, the Federal Government also withdrew funding for a new wind farm at Denmark in Western Australia, citing community opposition. The wind energy industry in Australia had been sent a very clear message. As a result wind energy companies in Australia are now looking to make their investments overseas, including China, and international interest in Australia has collapsed.

The wind farm controversy revealed some strange politics swirling around climate change. Local opposition to wind farms is heavily influenced by a network of anti-environmental activists with names such as Coastal Guardians and Landscape Guardians, some with links to the fossil-fuel and nuclear industries, and similarly named organisations abroad. This helps to explain why apparently independent local opposition groups reproduce the same misinformation and distortions about wind power. The campaigns around Bald Hills and others such as that at Bungendore in New South Wales are built mostly on a wave of disinformation aimed at bamboozling affected communities and crowding out legitimate debate about the pros and cons of wind energy.

Most opponents of wind farms seem to have little understanding of the threat posed to their local areas – let alone the entire globe – by climate change. While expressing concern about the environment, in campaigning against wind farms they seem unaware of the far greater threats to the local environment looming on the horizon. This has also proved true of established organisations that should know better. For instance, in April 2002 the Victorian branch of the National Trust of Australia organised a 'Wind power forum' at Deakin University. The event was staged in order to cultivate opposition to the development of wind farms and included speakers who were not just anti-wind campaigners but were also hostile to environmentalism as such. In particular, the National Trust gave a platform to Alan Moran of the Institute of Public Affairs, who had for more than a decade been one of the principal voices of anti-environmentalism in Australia. As we have seen, the institute has close links with a number of industry-funded denialist organisations in the United States and has been one of the more important centres for anti–climate change activism in this country.

The move against the development of wind power by the National Trust indicated that, even in 2002, it had virtually no understanding of the threat to Australia's natural environment posed by climate change. Its backward position may have been influenced by the fact that a number of its senior people are prominent members of the Liberal Party and were influenced by Canberra. Apart from anything else, this stance jeopardises decades of work by the National Trust aimed at conserving Australia's natural heritage. In the United Kingdom, anti-wind campaigners have made much of the threat turbines present to birdlife, but their claims have been spiked by the support of the Royal Society for the Protection of Birds for wind power and its identification of climate change as 'the most serious threat to wildlife'.

The politics of wind in Victoria have become Byzantine. While the federal environment minister insisted he supported the development of wind farms, his National Party ministerial colleague Peter McGauran denounced them as a 'complete fraud' that 'only exists on taxpayer subsidies'.[8] The endorsed Greens candidate for the Victorian state seat of Gippsland South, Jackie Dargaville, expressed antipathy to the Bald Hills wind farm on the grounds that there was 'community opposition'.[9] Here she lined up with Victorian Liberal leader Ted Baillieu, who took a populist stance that backfired.

In November 2006 the Federal Government, under intense pressure over its refusal to act on climate change, announced funding for a large solar electricity plant under its Low Emission Technology Demonstration Fund. Treasurer Peter Costello made the announcement with Steve Bracks and, by his presence, effectively endorsed the state Labor government's scheme to promote wind and solar. On the other side of the state in the seat of South-West Coast, Mike Noske, a former Liberal Party candidate and mayor, ran a pro-wind campaign against the Liberal member Denis Napthine, who, like his party, is strongly anti-wind.[10] The electorate hosts several wind farms.

GLOBAL CRINGE

The Eleventh Conference of the Parties held in Montreal in November 2005 confirmed, if confirmation were needed, that the role of the environment minister in the Howard Government is not to develop and implement measures to protect Australia's natural environment but rather to cover up inaction. Events at Montreal suggested that the minister lived in a different world from the one occupied by the rest of the delegates. As the nations of the world gathered to begin discussions for the second commitment period of the Kyoto Protocol, Campbell declared that Kyoto was dead and would not get beyond 2012. The debate had 'moved on', he said, and there was no need for Australia to ratify. Moreover, he claimed that other countries were realising that Australia was right not to join the protocol, and predicted the system for setting targets and timetables for greenhouse gas reductions could be scrapped after 2012. 'A number of [countries] are saying, "Look we made a mistake. We don't think that it's worth opening up a new negotiation about a future commitment when the commitments we have today are looking so unreasonable".'[11] He claimed that ministers from several other countries had told him: 'Australia made the right decision. This thing is not going to work'. Everyone in the vast conference centre knew that Australia had been the subject of unrelenting criticism from the international community at least since the protocol was agreed at Kyoto and especially since the Prime Minister announced in June 2002 that Australia would not ratify it. Yet in a unique interpretation of what it means to repudiate an international treaty, Campbell declared that 'no-one has shown more support for the Kyoto Protocol than Australia'.[12] That did not prevent him from attacking it at every other turn. According to Campbell: 'Signing Kyoto is like catching the three p.m. train from Central Station when it's five o'clock'.[13] The minister's confusion about departure procedures was mirrored in his understanding of international politics. Even when the Montreal meeting made unprecedented progress on the second commitment period after 2012, and was hailed around the world for injecting new life into the protocol, Campbell did not miss a beat.

His commentary on international climate negotiations has been reminiscent of the performance of the former Iraqi information minister Mohammed Saeed al-Sahaf. On 7 April 2003, as US tanks patrolled the streets of Baghdad only a few hundred metres from the location of his press conference, al-Sahaf claimed that there were no US troops in Baghdad and that the Americans were committing suicide by the hundreds at the city's gates. In his last public appearance as information minister the next day, 'Comical Ali' declared that the Americans 'are going to surrender or be burned in their tanks. They will surrender, it is they who will surrender'.[14]

While the rest of the world saw Australia as a pariah on climate change, Campbell proudly told all who would listen at Montreal that we were actually regarded as a world leader.[15] While the rest of the world thought that Australia was doing all it could to undermine Kyoto and to wreck any consensus, Campbell said that we were seen as a constructive player: 'Australia is in the middle of world efforts to defeat climate change. Australians should be very proud, not only of our domestic programs, but the fact that we are an integral and constructive player in all of the international fora'.[16]

Journalists naturally asked why Australia refused to adopt greenhouse gas reduction targets. Campbell reasoned: 'Targets are a proxy for doing the hard work'. In Comical Ian's world, accepting a legally binding limit on our greenhouse gas emissions is a substitute for undertaking real actions, presumably such as entering into another voluntary program with big business.

The same logic was on display in Campbell's reaction to a surprise judgment in November 2006 by the NSW Land and Environment Court. The court ruled that the NSW government's assessment of a proposed new coalmine at Anvil Hill should have considered the potential damage caused by the mine's greenhouse gas emissions. Senator Campbell reacted by declaring that the decision was 'disastrous' for the environment.[17] In Campbell's world of transcendental logic, burning fossil fuels is not the cause of climate change; it is the solution to it. Perhaps this astonishing confusion owes something to the insights of Alexander Downer. When asked why his Government would not impose limits on Australia's emissions – in the same way that the United States enforces limits on sulphur dioxide and every Australian state regulates urban air pollutants – he snapped: 'We are not trying to run some kind of police state'.[18]

Just before Montreal, the UN released a report showing large increases in Australia's greenhouse gas emissions. Campbell responded by saying that the UN report only told part of the story. He said: 'It's basically adding up all of the increases in the emissions but it isn't taking into account any of the measures we're doing to reduce emissions, so it's only part of the equation'.[19] Of course, if his policies were having any effect, the adding up would have given a lower number. In Campbell's world, the laws of arithmetic can be turned on their head.

Asked if we should be concerned about the new report that showed Australia's energy emissions rising rapidly, Campbell said: 'No, I think Australians should not be concerned about that, because we are a growing economy. We've got a growing population'.[20] In Campbell's world, if our greenhouse gases are growing because of population growth, then they do not count.

CAMPBELL AS CANUTE

Late one day in May 2006, the Federal Government posted on its website a report on the science of climate change commissioned from Professor Will Steffen of the Australian National University.[21] The purpose of the report was to provide a review of developments in climate science since the publication of the IPCC's Third Assessment Report in 2001. The Steffen report concluded that the upper estimate for global warming made by the IPCC, a global average temperature rise of between 1.4ºC and 5.8ºC by the end of this century, now appeared more likely to be reached or exceeded, and that observational evidence supporting the existence of climate change had grown even stronger in the five years since the Third Assessment Report. Among the indicators of a warming planet, Professor Steffen naturally included an assessment of the latest evidence of sea-level rise, which, he wrote, had since 1993 increased to about three millimetres per year.

Ian Campbell issued a media release to accompany the posting of the report. The Howard Government had again been caught out by a scientific assessment that rang alarms bells about climate change and implicitly highlighted the failure of the Australian Government to respond to the unfolding crisis. What sort of spin could be put on the report's release? It seemed impossible to deny the science, so the minister tried the customary tactic of simultaneously denying responsibility while claiming to take the matter very seriously. He said: 'Climate change is ... too serious a matter for Australians to be misled into believing that massive cuts to Australian greenhouse gases on our own will have any effect on global climate change'.[22]

So, according to the minister, climate change is too serious a matter for Australia to cut its greenhouse gas emissions. His crude attempt to deflect attention from the Steffen report did not stop *The West Australian* newspaper from running a major story on 26 May on the implications of sea-level rise for Western Australia, the senator's home state.[23] It quoted Professor Steffen along with Dr John Church, a senior CSIRO oceans researcher, who pointed out that forecast sea-level rise would result in the inundation of coastal land, including some expensive real estate. An oceanographer from the University of Western Australia, Professor Chari Pattiaratchi, was also quoted as saying that parts of the Perth foreshore, and popular seaside locations such as Mandurah, Australind, Dunsborough and Busselton, would be the first areas inundated by rising seas. Western Australia is particularly prone to sea-level rise because of the rapid warming of the Indian Ocean.

Campbell, who before entering parliament was a Perth real estate agent, took umbrage and launched an attack on the scientists and the science of climate change, branding their warnings 'ludicrous'. Perhaps imagining he had a special knowledge of coastal real estate, he seemed particularly offended at any suggestion that sea-level rise would affect the value of coastal properties and, in a surprising display of ignorance about the science of climate change, claimed that sea-level rise would not occur for 1000 to 2000 years. 'Climate change is a very serious issue', he said. 'However, we have trouble enough ensuring people take it seriously without ludicrous claims like this, i.e. people shouldn't invest in coastal properties.'[24] The newspaper followed up the next day with the headline: 'Science is debatable in the world of Senator Campbell'.[25]

It is unclear why Campbell made this extraordinary intervention, although *The West Australian* newspaper began investigating whether he owned coastal properties whose value might be affected by credible claims of sea-level rise. The minister's knee-jerk response is consistent with a pattern displayed by the Federal Government of formally accepting that global warming is occurring while denying any aspect of the science that seems politically unacceptable.

INSIDE THE TENT

In 2005 the Howard Government began to advance a new argument for its refusal to join international efforts to tackle climate change – in one word: China. Suddenly, China's booming economy was fingered as the real culprit in the global greenhouse debate. There is no point in Australia cutting its emissions, went the new mantra, when China's growth will swamp anything we do. A new figure was bandied about. Howard told his favourite broadcaster, Alan Jones, that 'if we closed down every power station in Australia tomorrow, it would take China nine months of emissions to cancel that out. Nine months'.[26] By February 2007 Malcolm Turnbull had cut the figure to five or six months.[27] With this argument, the Government was implicitly vilifying China as an irresponsible polluter, well aware of the historical use in Australia of subliminal fear of the yellow peril. It was Campbell who once again gave the game away.

In November 2006 Campbell travelled to Nairobi for the Twelfth Conference of the Parties. Campbell's media release on his departure asserted that China would become the world's biggest emitter of greenhouse gases by 2010 and that the Kyoto Protocol did not ask China or India to 'do a single thing' to reduce their emissions. Along with its transparent effort to cast China as the villain, the media release was also inaccurate. While not having a mandatory target, China took on a number of obligations under the protocol, including systematic reporting of its emissions and participating in the Clean Development Mechanism, which has led to a flood of major emission reduction investments. Moreover, all delegates at Nairobi other than Campbell were well aware that the Chinese

Government had gone much further than Australia in adopting policies to curtail its emissions. As a senior member of the UN secretariat pointed out, China is extremely touchy about any suggestion that it is not taking important steps to address climate change. In particular it has committed to sourcing 10 per cent of its energy from renewables by 2010, 18 per cent by 2020 and 30 per cent by 2030.[28] These figures make Australia's own renewables policy, now exhausted, appear trivial.

The Chinese delegation was seen discussing Campbell's release,[29] after which a senior Chinese official addressed the conference with a number of terse comments directed at Australia, criticising information he had seen 'that accused China of risking becoming the biggest polluter in the world'.[30] He pointedly noted that China's population is 65 times bigger than Australia's and drew the world's attention to the gulf between per capita emissions in the two countries. In the world of climate diplomacy, to publicly single out a country in this way is highly unusual and was a sharp rebuke. Australian officials were forced to build bridges knocked down by their bumbling minister. Pressure on Australia and the United States was increased when UN secretary-general Kofi Annan referred to the 'frightening lack of leadership' in fighting global warming.

Campbell's media releases provided entertainment for delegates at Nairobi with the extraordinary spin they put on Australia's role in the international climate change debate. Lauding the 'New Kyoto', a proposal which left senior UN officials scratching their heads and asking representatives of environment groups if they knew anything about it, the minister dismissed the Kyoto Protocol as a 'slogan', claiming Australia was 'interested in real solutions that will make a real difference'. Apparently oblivious to the perception of Australia as a persistent spoiler, Campbell said it was 'imperative the world worked together on climate change', reminding delegates that 'we all live in the same world'.[31] Australians cringed.

While the Australian Government continued to criticise the Kyoto Protocol and forecast its demise, behind the scenes it was making an extraordinary effort to continue to participate in the negotiations. The anxiety about keeping a seat at the table followed an incident in Montreal the previous November when UN security had been called to remove US delegates from a contact group of the parties. Then at a meeting in Bonn in May 2006, some parties to the protocol objected to the presence of Australia in the room. The Australian Government was humiliated at being treated as a mere observer; its refusal to ratify the protocol gave it the same status as NGOs such as Greenpeace and the nuclear lobby. Australia lodged a submission to the secretariat of the Framework Convention with legal arguments claiming that Australia had 'an express legal right to participate in all Protocol processes'.[32] Australia was particularly eager to participate in the vital Ad Hoc Working Group on Future Commitments, which had the task of negotiating the structure of the Kyoto Protocol in its second commitment period beginning in 2013. The exclusion of Australia from this group exposed the folly of the Government's position. At Kyoto it won an extraordinarily generous deal for the first commitment period, so generous that Australia needed to do nothing to meet its target, yet denied itself any influence over negotiations towards the second commitment period.

The European Union and the Framework Convention secretariat were unmoved by the submission, believing it had no legal basis. One of Australia's foremost authorities on international law, Professor Don Rothwell, noted:

> It would be arguable that a State which has clearly indicated that it does not intend to proceed to ratification or accession to the protocol and which has actively denounced the protocol should not continue to enjoy observer status when that State may be seeking to actively undermine the protocol.[33]

According to Rothwell, even the observer status of Australia was in question.

The Howard Government has refused to take on any obligation to cut Australia's emissions, and it also appears determined to prevent other nations from cutting theirs. To adapt the familiar metaphor, Australia wants to be inside the tent pissing in.

When the Bureau of Meteorology released figures in January 2006 showing that 2005 was the hottest year on record, Campbell declared: 'It's the hottest year, the hottest decade, the hottest minimum and the hottest maximum', before adding: 'The main thing is not to alarm people'.[34] We laugh, but in his own inept way Campbell was revealing his riding instructions from the Prime Minister: his job was to concede that climate change is real but persuade Australians that there is nothing to be alarmed about, so we should just relax. Unlike Robert Hill, who was never duped by his own spin, Ian Campbell seemed to suffer from a form of Stockholm syndrome; he was captured by and fell in love with his lies, so he could no longer distinguish between fiction and the truth. The transition from the sophistication of Robert Hill through the doughtiness of David Kemp to the slapstick foolishness of Ian Campbell indicates that Prime Minister Howard has not only continued to reject the science but has also wilfully thumbed his nose at it.

In 2000 Howard appointed Wilson 'Ironbar' Tuckey minister for forests. When Western Australian Liberal premier Richard Court was asked why Howard had made the appointment, he said that it was because the Prime Minister had a sense of humour. Perhaps the Prime Minister was also having a joke with his appointment of Ian Campbell as environment minister. Whatever the case, Campbell was replaced by a more formidable figure, Malcolm Turnbull, in January 2007, and was forced to resign from the ministry in March after being caught up in a political storm around disgraced former West Australian premier Brian Burke.

13. NEW GLOBAL MOMENTUM

TIPPING POINTS

The year 2005 was a watershed in the climate crisis, quite apart from the entry into force of the Kyoto Protocol. After several dispiriting years of slow progress, perceptions changed in response to a number of factors. There was the changing weather. The year 2005 proved to be the hottest on record, both globally and in Australia. The hurricane season in North America was the most intense and active in recorded history. Katrina and Wilma were particularly destructive, and by the time Katrina swept across New Orleans, some hurricane experts were willing to attribute the greater intensity of hurricanes to ocean warming due to climate change. Such developments were alarming enough, but they were also accompanied by a shift in climate change science. Previously, it had been thought that as greenhouse gas concentrations in the atmosphere increased, warming and various other changes in global weather patterns would gradually follow. Now scientists began to talk of 'abrupt' changes to global and regional weather patterns. This was the hypothesis of the Pentagon report on the security implications of climate change. The prospect of a stalling of the thermohaline circulation – the great ocean currents that take warm water from the tropics to the seas off the north-west coast of Europe – led to alarming scenarios such as the one depicted in the film *The day after tomorrow*.

Scientists began to talk about 'tipping points' to capture the idea that there might be hidden thresholds beyond which human-induced climate change would become drastically worse. In the words of one science writer: 'A tipping point usually means the moment at which internal dynamics start to propel a change previously driven by external forces'.[1]

Most of the concern about these danger zones relates to the crucial but little-understood role of the Arctic region.[2] Scientists have observed that the Arctic sea ice has been receding in summer more than usual. The exposure of dark seawater means that more of the sun's heat is absorbed, making it harder for the sea to freeze over again when winter arrives. Thus in 2005 there was 20 per cent less ice cover than over the period of 1979 to 2000, exposing an area the size of France, Germany and the United Kingdom combined that had previously been covered by ice. Some research suggests the process has passed a tipping point and there is no way back. If greenhouse emissions are not reduced, soon there may well be no Arctic ice cap. In itself the global warming effect of this may not be immense because less than 5 per cent of the Earth's surface is above the Arctic Circle. However, a dark North Pole could have dramatic effects on global wind patterns, including the jet stream, with profound implications for weather patterns everywhere.

The second tipping point to attract attention has been the possibility of the melting of the Greenland ice sheet. Although it may take centuries, this would raise sea levels by seven metres. All the indications are that Greenland is indeed melting. Modelling suggests that all of Greenland would melt, a 'climatic point of no return', if the average global temperature rose by 3.1°C, a figure in the middle of the range foreseen by the IPCC.[3] In a dramatic sequence in Al Gore's *An inconvenient truth*, torrents of warm water are shown rushing through cracks in the Greenland ice sheet, hastening the melting process. There are also concerns about the melting of the West Antarctic ice sheet.

This leads to the third tipping point – the melting of the Greenland ice sheet could pour masses of fresh water into the north Atlantic. This could prevent cold salt water from sinking and thereby close down the thermohaline circulation. Although the cooling effect of shutting down the Gulf Stream (part of the thermohaline circulation) in northern Europe is often overstated, the stalling of these great currents could have a dramatic effect on rainfall patterns in the Asian tropics, with far-reaching effects on food supply.

These are not the only tipping points causing alarm among scientists, some of whom now fear that they have seriously underestimated the likely extent and impact of climate change.[4] There is evidence that the permafrost covering vast areas of Russia and Canada is melting. This changes the albedo (light-reflecting property) of the surface, with the resulting warming leading to emission of potentially huge amounts of methane and carbon dioxide. There is also the effect of 'global dimming' to consider. Warming due to the enhanced greenhouse effect is to some extent offset by increased atmospheric concentrations of aerosols (mostly sulphur particles), also from the burning of fossil fuels. As measures are taken to reduce air pollution (in China and India, for example), the reduction in aerosols will perversely lead to greater global warming, especially in the northern hemisphere. Moreover, climate change may cause widespread die-off of forests, especially the Amazon, leading to a positive feedback effect as more carbon dioxide shifts from the terrestrial biosphere into the atmosphere, exacerbating global warming.

On the first day of 2007 Dr James Hansen, director of the NASA Goddard Institute for Space Studies and one of the world's most respected climate scientists, drew out the implications of these positive feedbacks in a terrifying interview.[5] He warned that we have less than ten years to cut carbon emissions before global warming runs out of control and changes the world forever:

> We just cannot burn all the fossil fuels in the ground. If we do, we will end up with a different planet. I mean a planet with no ice in the Arctic, and a planet where warming is so large that it's going to have a large effect in terms of sea-level rises and the extinction of species.

Noting that half the people in the world live within 15 miles of the coast, he said, 'The last time it was 3°C warmer, sea levels were 25 metres higher … once you get the process started and well on the way, it's impossible to prevent it'.

Partly in response to these disturbing scientific developments, in June 2005 the national science academies of the G8 nations, along with those of Brazil, China and India, signed a joint statement explicitly endorsing the IPCC consensus on climate science. The statement declared: 'The scientific understanding of climate change is now sufficiently clear to justify nations taking prompt action'.[6] This unique joint endorsement of global scientific opinion was directed unapologetically at the intransigence of the Bush Administration.

MOVEMENT IN THE US CONGRESS

With scientists around the world sounding alarms, segments of the US population were motivated to act. They faced formidable obstacles. The Bush Administration remained firmly in the grip of the more recalcitrant elements of the fossil-fuel lobby and had an ideological opposition to taking action on climate change. One of Bush's first acts as President had been to repudiate his election promise to cap carbon dioxide emissions from US power plants. When asked in 2001 if the President would be urging Americans to restrain their energy use, Bush spokesman Ari Fleischer replied: 'That's a big "no"'. He went on to declare that wasting energy is akin to godliness:

The President believes that it's an American way of life, that it should be the goal of policy-makers to protect the American way of life. The American way of life is a blessed one ... The President considers that Americans' heavy use of energy is a reflection of the strength of our economy, of the way of life that the American people have come to enjoy.[7]

Bush's fundamentalism soon proved out of step not only with the science but also with mainstream opinion in the United States, opinion that has strengthened throughout the presidency to the point where Bush is isolated on the issue in his own land as well as internationally.

In October 2003, a group of powerful Republican and Democrat senators introduced the McCain–Lieberman Bill to the US Senate, which would have required major emitters in the United States to adhere to mandatory, economy-wide emission caps. The fact that the bill was introduced at all represented a radical break from the view that the United States is irredeemably opposed to tackling climate change. The bill proposed capping emissions at year 2000 levels over the period of 2010 to 2016, and in subsequent years reducing them to 1990 levels.[8] The caps would apply to all major emitters of carbon dioxide and other industrial greenhouse gases. Transport emissions would be tackled by requiring refineries and importers of petroleum to hold allowances for each tonne of carbon dioxide that would be emitted in the combustion of their products. The proposed program would have covered more than 70 per cent of all US carbon dioxide and industrial greenhouse gas emissions.

The Senate vote on the McCain–Lieberman Bill was lost 43 to 55. Observers were astonished at how narrow the margin was, and how close the United States had come to having a far-reaching bipartisan mandatory program to cut greenhouse gas emissions (although the bill would still have had to be passed in the lower house, a more difficult task). Some senators who voted against the bill nevertheless spoke in favour of it. They made it very clear that they knew the United States had to act soon and that they did not want to go down in history as the law-makers who prevented action being taken. The closeness of the vote anticipated 2005's global shift on climate change, and Washington observers believe that a similar bill will be introduced again before long. There is every likelihood it will become law. In April 2006 the head of General Electric, by some measures the world's biggest corporation, said that the very first act of the next US president will be to initiate urgent action on climate change.[9] In the same month, John McCain declared: 'Joe [Lieberman] and I will make the US Senate keep on voting and voting and voting and voting'.[10]

This does not mean that the United States is likely to ratify the Kyoto Protocol, and certainly not during the Bush presidency, but it does point strongly to serious reduction measures being implemented in the foreseeable future. Moreover, Senator John McCain is a front-runner as the Republican candidate for the presidency. Some analysts expect that a strong emission reduction regime in the United States would, over time, converge on the Kyoto regime of Europe and Japan.

In June 2005 the US Senate resolved 'that the United States act to reduce the health, environmental, economic and national security risks posed by global climate change'.[11] Despite what *The New York Times* called 'ferocious White House opposition', the Senate also voted in support of a program of mandatory emissions controls. The non-binding nature of the Senate resolution allowed opponents to dismiss it, but as *The New York Times* editorialised:

> The resolution was anything but meaningless. It represents a major turnaround in attitudes, especially among prominent

Republicans who only a few years ago doubted a problem even existed. It is something to build on: Pete Domenici, the most influential Senate Republican on energy matters and a recent convert to the global warming cause, has already scheduled hearings to see what sort of legislation can be devised down the road.

And it terrifies the White House because it is further proof that the administration's efforts to minimize the warming threat have failed and that President Bush's voluntary approach to the problem is no longer taken seriously.[12]

Another prominent Republican, Senator Olympia Snowe, followed up by declaring: 'As the US Senate officially recognised for the first time last week, there is no doubt that greenhouse gases are irrevocably impacting our climate and that mandatory caps on greenhouse gas emissions are necessary'.[13]

In December 2005 the first mandatory caps on US greenhouse gases were imposed by an alliance of seven north-eastern states. The scheme has been designed with an eye to future integration with the Kyoto trading system. In the same year, California Governor Arnold Schwarzenegger signed into law ambitious targets to cut his state's greenhouse gas emissions. Methods include renewable energy targets and caps on emissions from power plants. At the 2005 Montreal Conference of the Parties, the US delegation was embarrassed by representatives of the mayors of around 250 US cities who pledged to cut their emissions because of the failure of the Federal Government to act.

The world has been a hostage to the lack of synchronicity in the US political system. During the Clinton years and the negotiation of the Kyoto Protocol, the presidency was in advance of the legislature. During the Bush presidency, notably in his second term, the legislature has been ahead of the presidency. The world must wait until the two are aligned on the need for the United States to join global efforts to reduce greenhouse gas emissions.

Across the Atlantic, the business environment was transformed with the implementation on 1 January 2005 of the European emission trading system that caps greenhouse gas emissions across the European Union. It is the biggest and boldest pollution trading scheme ever conceived, and its success is critical to the future of international efforts to tackle climate change. The first, or 'warm-up', phase, applying to 25 European Union countries, will run until the end of 2007. The second phase will begin in 2008 and end in 2012, the first commitment period of the Kyoto Protocol.

Initially the European trading system will cover carbon dioxide emissions from four broad sectors: energy (including electricity), iron and steel, minerals, and pulp and paper. Each installation is allocated allowances by the relevant national government. It's estimated that more than 12,000 installations accounting for 46 per cent of total European Union carbon dioxide emissions will be covered. The plans must be approved by the European Commission to ensure that no nation gives any sector an easy ride that may allow it to undercut competitors in other countries. This is something of which Australia ought to take note: it shows how serious the European Union is about ensuring that firms required to cut emissions do not face unfair competition.

ASIA-PACIFIC PARTNERSHIP

By injecting new life into the Kyoto Protocol and marginalising the United States, the 2005 Montreal conference presented the Australian Government with a big problem. Scorching temperatures across eastern Australia in December 2005 and January 2006 highlighted the Coalition's continuing intransigence. The Bureau of Meteorology had recently reported a 40-year decline in rainfall across southern Australia, with precipitation declining by 25 per cent in some areas. Many dams supplying cities and towns around the country remained at record low levels. Munich Re, one of the world's largest insurance companies, announced in late December that economic losses due to weather-related natural disasters set a new record in 2005, reaching more than US$200 billion.[14]

However, in response to continuing public pressure, the Government had acquired a new trump card that it was now ready to play. Having rejected Kyoto, and needing to appear to be doing something, the Government threw its weight behind a new agreement, the Asia-Pacific Partnership on Clean Development and Climate, soon known as AP6, which brought together Australia, the United States, Japan, South Korea, China and India. The initiative for this had come from the Bush Administration; the Howard Government, seeing the opportunity to deflect criticism that it refused to join international efforts, needed no persuasion to get on board. China, India and South Korea needed some coaxing, but faced with promises of financial transfers to fund clean technology investments and promises that the new partnership would not jeopardise the Kyoto treaty, which the three had signed and ratified, they agreed to participate. Japan was much more reluctant, suspecting that it was a plot to protect the United States and Australia from accusations of being internationally irresponsible. Under pressure from the Bush Administration, Tokyo relented at the last moment. It was not an auspicious start.

Prime Minister Howard described the pact as a 'historic agreement for the cause of reducing greenhouse gas emissions', which 'establishes a long-term framework for co-operation among key countries in the region'.[15] Minister Campbell, using the language of environmentalists that made his more hard-headed colleagues blanch, referred to the pact as 'an historic breakthrough on saving the climate and saving the planet'.[16] But someone had forgotten to explain to the enthusiastic environment minister what this was all about. When the partnership was announced in July 2005, he let the cat out of the bag and embarrassed the other members by blurting out that AP6 was an 'alternative' to Kyoto. He was quickly corrected by wiser heads, who insisted that it was but a *complement* to the Kyoto Protocol; after all, four of the six members had ratified the latter. Asked about Campbell's comments in Montreal in November 2005, Harlan Watson, the head of the US delegation to climate meetings and legendary hard man, gently rebuked Campbell for his gaffe.

Significantly, none of the delegates from China, India, Japan and South Korea mentioned the Asia-Pacific Partnership in their high-level addresses to the plenary of the Montreal conference, and all four strongly affirmed their continued commitment to the UN process. Informally, they have indicated that they are willing to participate in the partnership for as long as it generates a substantial program of activity and is not another smokescreen to cover inaction on the part of the United States and Australia. These countries made it clear that the new partnership must be entirely in harmony with Kyoto and that it must not be used to undermine it. The Chinese Ministry of Foreign Affairs was unambiguous: 'This pact has no power for legal restrictions. It is a complement to the Kyoto treaty, not a replacement'.[17]

Nevertheless, the Australian Government cranked up its propaganda machine. *The Australian,* the first newspaper worldwide to announce the new pact in late July 2005, was as willing as ever to oblige on this count. The story was a big enough 'scoop' to occupy the page-one lead of the Saturday edition of the newspaper – prime space frequently reserved for Government propaganda – four weeks before the Sydney meeting of AP6. Under the headline 'Climate fund's $100 million kickstart', it began by saying the Government was 'considering' injecting $100 million into a fund to promote advanced energy technology in China and India. It was a pure puff piece, in which the author, Steve Lewis, wrote that neither China nor India 'has been prepared to ratify the Kyoto Protocol', a clanger that suggested that he had failed to check any of the information fed to him by the Government.[18]

In the same edition, an editorial attempted to dismiss the significance of record temperatures, calling 2005 not the 'hottest' year but the 'warmest' and saying that one year's statistics 'do not mean the sky is falling down'.[19] Warming to its theme, the Murdoch broadsheet attacked Greenpeace for exaggerating the likely impacts of climate change. Presumably with a straight face, the editor wrote that if Australia joined Kyoto, 'the economy would go into a tailspin'. Breaching perhaps the last journalistic standard still intact, a few days later the paper carried a major story that cast doubt on the science of climate change, talking of 'theories and counter-theories', pointing to the error of 'blaming industry', claiming that the cause of warming was 'still unproven', referring to the 'infant nature of the scientific debate', and so on.[20] What was remarkable about this piece was that it was not published as opinion but rather presented prominently as a news story. Moreover, there was no by-line; the news story was attributed to 'a special correspondent'. The paper felt compelled to add at the bottom: 'The correspondent is employed by a resources lobby'. The writer was almost certainly a coal industry PR man named Matthew Warren.

In the following days, the opinion editor ran the same old pieces from sceptics such as Ian Plimer, who claimed that a temperature rise of a 'few degrees' does not matter and (posing as an expert in economics) that 'only a strong economy can produce the well fed who have the luxury of espousing with religious fervour their uncosted, impractical, impoverishing policies'.[21] As is often the case, it was left to the cartoonist to expose the lies about AP6. Peter Nicholson drew a Chinese official saying to a disgruntled Howard: 'and here's some leading edge technology from *us* ... a pen to sign Kyoto'.

Although Government propaganda seemed to have persuaded some journalists otherwise, four of the six members of the Asia-Pacific Partnership – China, India, Japan and South Korea – had ratified the protocol and thus undertaken to fulfil a number of legally binding obligations under it. While the Government made much of the fact that the new pact included countries responsible for nearly half of global greenhouse gas emissions, the Kyoto Protocol accounts for 75 per cent of global emissions, and mandates action.

The inaugural meeting of AP6 was arranged for Adelaide in November 2005, but was postponed when US secretary of state Condoleezza Rice could not attend. It was rescheduled for Sydney in January, but as it turned out Rice still did not attend – a bad sign for the Government as it meant that public interest would be harder to generate. In another bad sign, US Republican senator John McCain, then (as now) tipped by many to be the next Republican presidential candidate, said that the Asia-Pacific Partnership 'amounts to nothing more than a nice little public-relations ploy ... It has almost no meaning. They aren't even committing money to the effort, much less enacting rules to reduce greenhouse gas emissions'.[22] David Sandalow, a former US state department official who is now at the Brookings Institution, said: 'It's a great line-up of countries; I just wish they were doing something serious. Basically these kind of technology-cooperation partnerships have been around for years. This seems to be nothing but a repackaging of existing technology partnerships tied up in a bow.'[23]

The first AP6 meeting failed to generate the positive press that the Government had hoped for, not least because the parties agreed to do almost nothing. A number of working groups were formed. Critics and supporters alike went home asking, 'What was that all about?' In his speech to the conference, Prime Minister Howard repeatedly used the word 'practical' to describe his response to climate change, contrasting it with 'unrealistic' approaches. This is a motif of his Government, which uses 'practical' as a way of differentiating itself from the 'theoretical' approach of its opponents. In reality the word signals that the Government plans to do nothing, as has been the case with both its 'practical approach to climate change' and its commitment to 'practical reconciliation'. Stressing to delegates that the partnership was certainly not a competitor to the Kyoto Protocol but 'more an augmenter', Howard told the delegates that an analysis by Australian Bureau of Agricultural and Resource Economics showed that AP6 had:

> the potential to reduce greenhouse gas emissions in partner countries by almost 20 per cent below what would otherwise be the case by the year 2050. The spillover to the rest of the world could lead to a global cumulative reduction in emissions of some 13 per cent below what would otherwise occur over the same period.[24]

When the ABARE report was released soon after, it was clear that the Government had shot itself in the foot. Although based on a set of unrealistic assumptions about the efficacy of voluntary actions, ABARE's modelling nevertheless concluded that under the best-case scenario, annual global emissions would increase from approximately eight gigatonnes of carbon equivalent at present to over 17 gigatonnes in 2050 under the influence of the AP6 agreement (see Figure 6, p. 192).[25] The consensus among climate scientists is that annual emissions must be reduced to around three gigatonnes to prevent the worst effects of global warming. So 14,000 million tonnes of carbon annually had gone missing in the Government's calculations. Having at various times criticised the Kyoto Protocol for not going far enough, its alternative to Kyoto would go much less far.

The Government did its best, under the eye of an increasingly sceptical media, to make AP6 look like a bold new plan to tackle

FIGURE 6: Global emissions projections for 2050 showing modelled AP6 results and the position recommended by scientists

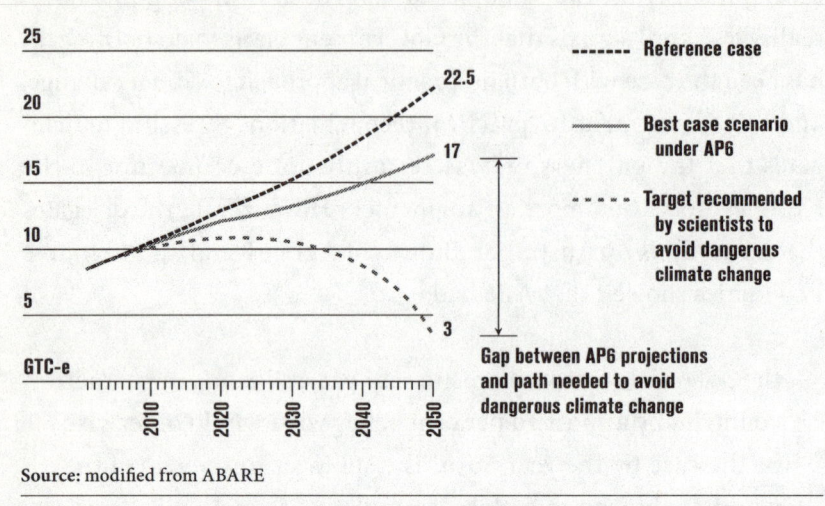

[Figure: Line graph showing GTC-e on y-axis (0–25) and years 2010–2050 on x-axis. Reference case rises to 22.5; Best case scenario under AP6 rises to 17; Target recommended by scientists to avoid dangerous climate change falls to 3. Arrow indicates the gap between AP6 projections and path needed to avoid dangerous climate change.]

Source: modified from ABARE

climate change. But the lukewarm commitment of the other parties made this a difficult task. Although it was noted that the United States made no financial commitments at the Sydney meeting, it was expected that the Bush Administration would secure funding through Congress. However, in late May 2006, the House Foreign Operations Appropriations Subcommittee refused a White House request for US$46 million to fund commitments under the Asia-Pacific Partnership, a refusal that was repeated a little later, effectively neutering the initiative. With uncharacteristic understatement James Connaughton, the head of the White House Council on Environmental Quality, said: 'If we don't get the budget, it will be a great challenge'. As it is fair to assume that India and China signed on to AP6 solely because they expected to receive some funding from the United States and Australia for energy projects, AP6 must now be looking much less attractive. But the failure of AP6 to give it the credibility it so desperately needed was only the opening salvo of what would be a disastrous year for the Howard Government.

14. THE TURNING TIDE

AN INCONVENIENT TRUTH

After a decade of spin, evasion and inaction, the Australian Government and the fossil-fuel lobby saw events turn dramatically against them in 2006. At the beginning of the year the AP6 conference in Sydney failed to give the Government the boost it had hoped for. It met with sceptical press coverage and public bemusement. The changing climate appeared to be having a worrying effect on daily life. The news that 2005 had been the hottest year on record was hard to explain away to the urban population, but it was the unrelenting drought that finally seemed to bring about a shift in the public mood. The winter rains failed and the situation became increasingly dire. Dams were not replenished and it began to dawn on city dwellers that water restrictions were here to stay.

In May *The Sunday Age* carried a special report on 'Life in the greenhouse' featuring a graphic mock-up of what the Melbourne central business district would look like if sea levels were to rise by six metres – the top end of projections under some climate scenarios. It quoted Professor Will Steffen, the eminent Australian National University scientist whose report on the implications of climate change for Australia had just been delivered to the Government, saying that a city under water 'is not out of the realm of possibilities'.[1] Any newspaper editor takes a risk in making such a big call, but *The Sunday Age* had decided that the evidence was now so strong that it could risk ridicule by climate sceptics. The same story quoted three celebrities – football coach Mick Malthouse, netballer Liz Ellis and television presenter Andrew O'Keefe – expressing alarm and criticising the Government for its inaction. Climate change had become mainstream with a vengeance.

The visit of Al Gore to Australia in September 2006 to promote his film *An inconvenient truth* was a pivotal moment in the debate. The film was the culmination of Gore's two-decade-long commitment to alerting the world to the perils of climate change. Like Churchill's warnings of the looming dangers of European fascism, Gore's message had for years fallen on deaf ears. Gore's whirlwind trip included a number of powerful media appearances that lent authority and urgency to the message of the film. He politely but firmly singled out Australia as the only nation to support the Bush Administration in its recalcitrant stance and suggested that if the Howard Government changed its position, it would put enormous pressure on the United States to do likewise.[2] In an age of visual media, the documentary had more impact than any number of newspaper articles and books.

For the Government, the preferred media strategy in such cases is to attempt to ignore the issue so as not to give it oxygen. But when events begin to dominate the news media, silence speaks volumes. The Government was either caught off-guard, or it simply could not compose an effective response to Gore. When industry minister Ian Macfarlane was asked about the film, the Government's foremost greenhouse troglodyte could not help himself. Gore was 'here to sell tickets to a movie' he growled on the ABC's *AM* program.[3] Perhaps, along with Senator Nick Minchin, the last elected official in Australia still willing to deny in public the science of global warming, he dismissed the film as 'just entertainment'. That of course was the problem for the Government – the issue most Australians would prefer to ignore had been turned into entertainment. The Prime

Minister tried to take the high ground, declaring that he did not take policy advice from films.[4] An enthusiastic Ian Campbell endorsed the science – declaring that 'my most respected scientists concur', as if he owned the CSIRO – but went on to dismiss the Kyoto Protocol once more as a 'slogan'. Rumours began circulating that the Prime Minister would be shifting Campbell out of his portfolio, and probably out of the ministry, with the next reshuffle.

Howard showed again that, despite the massive evidence and virtual unanimity of expert opinion, he still did not accept that global warming was occurring, declaring that he was 'sceptical' about the predictions. When asked to respond, Gore was scathing: 'There's no longer debate over whether the earth is round or flat, though there are some few people who still think it's flat'.[5] The contrast between Howard and the phalanx of business leaders interviewed on a *Four Corners* program in August 2006 was stark. After more than a decade in office and a mountain of scientific reports from every credible source in the world, in response to a question about the need for deep cuts in greenhouse gas emissions Howard could still say: 'Well, I want to see the evidence, I want to see the science'.[6] In truth, he did not want to see the evidence, which had been accumulating on his desk for years.

The intense media interest in Gore's visit exposed the poor quality of the greenhouse debate in Australia. A few days before Gore arrived, *An inconvenient truth* was reviewed by Margaret Pomeranz on the ABC TV's *At the movies*. Giving the film four stars, Pomeranz wrung her hands, saying, 'Surely it's possible to set up a body that doesn't have any vested interest in either outcome ... [one] that just wants to know the truth'.[7] There could be no better illustration of the effectiveness of the fossil-fuel lobby's campaign to confuse the public by creating the impression that the scientists cannot agree. The gaffe was all the more remarkable because Pomeranz, an otherwise sympathetic and intelligent commentator, had just watched Gore explain that over the previous ten years, in a random sample of 928 peer-reviewed articles dealing with climate change in professional journals, not one had called into doubt the causes of global warming. There was no scientific disagreement of any note. Yet journalists and commentators had been manipulated by the fossil-fuel lobby into believing that there were 'two sides' to the question and that they had therefore, in the interests of balance, to present both.

Pomeranz is not a political journalist and can be forgiven for her poor understanding of the issue. No such excuse was available to Fran Kelly, the presenter of ABC Radio National's influential breakfast program. As Kelly challenged Gore on the science and the politics of climate change, it was as if she were interviewing the chief medical officer and aggressively claiming that 'some doctors' argue that smoking does not, in fact, cause cancer.

MURDOCH AND HIS PAPER

The most intriguing reaction to the Gore visit and his film came from the Murdoch flagship *The Australian*. As we have seen, *The Australian* has run a virulently anti-greenhouse line for years, especially after Chris Mitchell took over as editor-in-chief in 2002. Mitchell was notorious among environmentalists in Queensland for his fanatical anti-green views while editor of *The Courier-Mail*, the Murdoch paper that monopolises Brisbane. From the outset at *The Australian*, Mitchell did not simply turn over editorial space and opinion pages to climate change sceptics and denialists, but he also allowed the news pages themselves to become a parody of dispassionate journalism, a phenomenon that reached a climax in the lead-up to Gore's visit. On 2 September, *The Australian*'s Saturday edition carried a page-one lead under the headline: 'Science tempers fears on climate change'. Claiming it had 'exclusive' access to the draft of the Fourth Assessment Report of the IPCC, the paper opened with the claim: 'The world's top climate scientists have cut their worst-case forecast for global warming over the next 100 years'. These were the same scientists *The Australian* had been bagging for years. As for the 'exclusive', the document in question had been posted on the internet by the US Government several weeks earlier. The story was a serious distortion of the draft IPCC report which, on the basis of much better climate modelling, had narrowed the scope of its warming projections – from 1.4°C to 5.8°C in the 2001 IPCC report to 2°C to 4.5°C in the current one. To suggest, as the story did, that we can breathe a sigh of relief and not worry too much about global warming was a travesty of the facts.

The verballing of the IPCC continued the next day in an editorial that claimed the world's scientists had rejected 'doomsday scenarios' by concluding that the planet is heating 'nowhere near as much as once thought or feared'.[8] The rest of the editorial read as if it were written by the Prime Minister's press secretary. It ran through all of the worn-out arguments, referring to natural variability and the little ice age, the mistaken 'hockey stick' graph, winners as well as losers from climate change, Australia's 1 per cent contribution to global emissions, real wages to be cut by 20 per cent, the answer is in clean coal, on track to meet our Kyoto target, AP6 is the answer, and on and on. But the true agenda of the newspaper's editor was revealed in the following:

> The [IPCC] report is particularly valuable as a rebuke to that radical and disproportionately loud fringe of greenies and leftists who treat environmentalism as religion and for whom humanity's sinful, decadent ways threaten to bring down the wrath of nature or the gods and must be changed.

For *The Australian*, it appears, science is merely a tool to be deployed in a larger ideological battle, one it is waging against an imaginary enemy. The news that the Cold War had ended in 1989 has yet to filter through to its editors and columnists.

The 'scoop' trumpeted by *The Australian* was written by the newspaper's new environment reporter, Matthew Warren, a name that was unfamiliar to close watchers of the climate change debate. In August the paper had announced Warren's appointment by writing that he had 'spent the past 12 years working in environmental policy and strategy, both in Australia and overseas. He has advised industry and governments on a range of technical and strategic environmental policy issues'.[9] What *The Australian* did not say was that Warren's previous job was director of external affairs for the NSW Minerals Council. As the PR flak for one of Australia's most active coal lobby groups, Warren was responsible for media campaigns attacking green groups protesting about climate change. Apparently without irony he subsequently told *Crikey* that he had joined *The Australian* 'out of growing frustration about the quality of the public environment debate in the media'.[10]

With this background, it was reasonable to expect *The Australian* to launch a campaign of ridicule and vilification against Al Gore and *An inconvenient truth*. However, in the two weeks before Gore's visit there had been a seismic shift in the Murdoch empire's position on global warming. At the invitation of Rupert Murdoch, Gore had addressed the annual July gathering of News Corporation executives at Pebble Beach, California. They watched *An inconvenient truth*. This was a major breakthrough, as Murdoch's news outlets, led by the virulently anti-green Fox TV, had for years run a hostile campaign against action on climate change. Murdoch executives listened attentively to Gore, although the mood was apparently soured by an embarrassing interjection from the floor by Andrew Bolt, the bumptious right-wing 'attack dog' from *The Herald Sun* in Melbourne. Gore is said to have made mincemeat of him. It is unclear what induced the switch in the position of the Murdoch empire, although James Murdoch, Rupert's son and heir to the empire after the departure of Lachlan, is said to be very pro-green, even to the point of making his UK pay television station BSkyB carbon-neutral.[11]

Murdoch's defection from the denialist camp had an instantaneous effect around the world. *The Sun*, the London tabloid said to be Murdoch speaking without the quote marks,[12] and whose political editor, Trevor Kavanagh, has been called Tony Blair's real minister for Europe, spilled onto the streets on 13 September calling on its readers to 'Go green with the Sun'. The edition, and following ones, were chock-full of stories about the dangers of global warming and the need for urgent action. Under a huge headline: 'Man the lifeboats' and a graphic of a sea-encroached British Isles, it asked: 'Will your town be under water if global warming takes hold?' Twenty articles on climate change followed, including an editorial declaring, 'Too many of us have spent too long in denial over the threat from global warming. The evidence is now irresistible'. After the page-three topless babe and the 'celeb directory', Kavanagh weighed in with: 'Only a fool would turn his face against the evidence of this summer – the second record-breaker in three years'. The campaign continued for day after day, with new and surprising angles being found by the tabloid. It did not take long for it to link climate change with its most popular theme. 'Going green can be so sexy', *The Sun* declared, next to a picture of 'Ruth', topless and with pink hot pants straddling a bicycle (men's, of course).[13]

But if, after the Sun King's epiphany, *The Sun* was rooting for action on climate change, someone forgot to tell Chris Mitchell at *The Australian* in Sydney. The paper ramped up its attacks, especially around the time of Gore's visit. Kicking off with a front-page banner reading 'Why Gore is wrong', it editorialised in favour of everything the Government had said. The opinion editor Tom Switzer commissioned a series of articles from the usual suspects.

Sceptic William Kininmonth wrote that 'computer models are not reality and alarmist predictions have no sound basis'. An unknown named Dean Bertram characterised Gore's film as science fiction and advanced the latest piece of sceptic pseudo-science, the claim that Mars is getting hotter too, so global warming must be due to sun spots. He praised Margaret Pomeranz for calling for the scientists to get together and sort out the facts. Anne Henderson of the right-wing Sydney Institute displayed the growing desperation of the anti-greenhouse brigade in a nasty piece arguing that, despite his moral elevation, Al Gore was personally responsible for a lot of greenhouse gas emissions and that 'there is no evidence that Gore has signed up to use green energy'.[14] Well, there is 'no evidence' that Anne Henderson does not dump her rubbish in Sydney Harbour. For good measure, the former coal lobbyist turned independent journalist Matthew Warren reviewed *An inconvenient truth*, calling it 'a horror film with a contrived plot'.[15] He quoted tireless sceptic Bob Carter saying the film was 'pathetic'.[16] Clearly echoing Exxon-funded websites, Warren wrote: 'Despite nearly two decades of consolidated research on the subject, there is still limited agreement about climate-change science'. Perhaps foreshadowing the need to recast the sceptics as noble, if ultimately wrong, seekers of the truth, Warren referred to Karl Popper's falsifiability principle, arguing, 'Only by continually challenging scientific theories do we make progress'. The sceptics could be wrong, but what an invaluable part they were playing in the process of finally getting to the truth. Warren followed up with a series of anti-greenhouse articles, including one that mocked the decision by the Australian Football League in Melbourne to reduce the greenhouse emissions associated with football matches.[17]

The Australian was now virtually alone among major newspapers around the world in maintaining its denialist stance on climate change. *The Times*, also owned by Murdoch, had switched. The influential London magazine *The Economist*, which had heavily promoted Bjorn Lomborg's work and ridiculed environmentalists, shifted position, arguing that the climate system could 'spin out of control', that 'the slice of global output that would have to be spent to control emissions is probably not huge' and urging Bush to take climate change seriously.[18] In Australia, *Business Review Weekly* declared: 'As long as Australia continues to refuse to ratify the Kyoto protocol it continues to fail the private sector by depriving it of the key mechanism any market needs to spark innovation – a price signal'.[19] Besides *The Australian*, the only other major hold-out was *The Wall Street Journal*, which, unlike *The Australian*, has for the most part reported the issue fairly in its news pages, but its editorials are stridently anti-greenhouse and rarely miss an opportunity to distort the truth. It is said that *The Wall Street Journal*'s reporters distance themselves from the paper's editorial line and laugh or cringe when they are challenged on it.[20] With a few notable exceptions, the journalists at *The Australian* show no such sensitivity when it comes to climate change.

There were occasional signs that *The Australian*'s resolve to remain the last bastion of climate change scepticism in the world's press was weakening. On 26 September it carried a cut-down version of an article by James Murdoch that had appeared in, of all places, *The Guardian*.[21] There the head of BSkyB criticised Exxon for funding sceptics groups and described the actions he had taken to make BSkyB the world's first carbon-neutral media company. The heir to the News Ltd throne finished with a flourish:

> Humanity is incredibly innovative. We have the capacity to solve the problem of climate change; the only issue is whether we as individuals, governments and businesses have the courage to act together to do what needs to be done. The stakes could not be higher.

James Murdoch was not the only unexpected source to target ExxonMobil. In September 2006 the Royal Society, Britain's leading scientific academy, took the highly unusual step of writing to Exxon asking that it desist from funding organisations that 'have misrepresented the science of climate change by outright denial of the evidence'.[22] The report mentioned the Competitive Enterprise Institute and the London-based International Policy Network. ExxonMobil's response was to sound wounded.[23]

At the same time two US senators, one from each of the main parties, wrote to Exxon with the same request – stop funding groups that claim global warming is a myth.[24] The letter came as the Union of Concerned Scientists completed a report showing that ExxonMobil had funded 29 climate change denial groups in 2004 alone. Since 1998 more than US$16 million had been spent by the corporation funding groups such as the Competitive Enterprise Institute and Tech Central Station.[25] In December a European watchdog identified the Centre for the New Europe and the International Policy Network as recipients of Exxon largesse. 'Covert funding for climate sceptics is deeply hypocritical because ExxonMobil spends major sums on advertising to present itself as an environmentally responsible company', wrote the Corporate Europe Observatory.[26]

Despite the continued efforts of Exxon, some of the most rabid sceptics began subtly to change their tune. Before Rupert's epiphany, Alan Wood, *The Australian*'s economics editor, wrote at least a dozen anti-greenhouse articles for the paper between 2003 and 2006 in which he attacked 'climate change scare-mongering' and wrote that 'a fog of green hysteria has descended over the global warming debate'.[27] In an especially ignorant remark, he asked: 'Do you think people who can't tell you whether it will rain next Wednesday are really capable of building models that tell you what the climate will be like 100 years from now?' Meteorologists may not be certain what the weather will be tomorrow, but they can tell us with a very high degree of certainty what the average daytime temperature will be in Melbourne next summer, and the summer after that.

Yet in October 2006 Wood was easing into a back-flip. After a sustained rant against 'environmental zealots', he changed his tune: 'Climate change is taking place, but how fast it will proceed, what its causes and consequences are, and what can, or should, be done to attempt to mitigate it are still matters of legitimate debate, not the subject of a phony scientific consensus'.[28]

However, by December Wood was back to his denialist ways. His stance is reminiscent of the Black Knight in the *Monty Python* skit: despite having both arms and legs chopped off, he shouts, 'Come on, I'll fight you'.

GOING NUCLEAR

John Howard has been thinking about nuclear power for a long time. It seems that at some point Liberal Party polling, or anecdotal evidence, started to show a softening of public attitudes towards uranium mining and export and the possibility of nuclear power itself. Howard has said that he believes the public mood has evolved. 'I think it's changed a lot from the early 1980s. I've been surprised by the number of environmentalists who have said they are prepared to look again at nuclear power as an energy source'.[29]

The usual method of testing a new and risky message is to send someone out to fly a kite. The task fell to defence minister Brendan Nelson, who in April 2005 gave a speech which urged serious consideration of nuclear energy in Australia. The Government had identified rising public concern over climate change as a justification for major new energy sources. Nelson chided the public for succumbing to scaremongering: 'Although much hysteria surrounds global warming, it pales into insignificance compared to that surrounding nuclear power'. Noting that, as the supplier of a fifth of the world's uranium, Australia is already part of the nuclear fuel cycle, Nelson appealed to both the financial and environment benefits of going nuclear.[30] Emboldened by the absence of public outrage, Nelson turned up the heat in a speech in August. In what were described as 'the most pro-nuclear remarks made yet by a member of the Howard Government', he said that it would be necessary for Australia to adopt nuclear power within 50 years in order to cut greenhouse gas emissions.[31]

Alexander Downer also became a kite-flyer for the Prime Minister. In a September 2005 speech he said: 'Nuclear power's clear benefits in greenhouse terms are causing many countries to reconsider some outdated prejudices. The reality is that nuclear energy is the only established non-fossil-fuel energy source capable of generating large amounts of baseload electricity without significant emissions of carbon dioxide'.[32] Safety concerns are exaggerated, he said, and then took a moral turn: 'And as the holder of the world's largest uranium reserves, we have a responsibility to supply clean energy to other countries'. No mention was made of our responsibility for the greenhouse gases associated with our coal exports. He ended by pointing to the 'reality' that nuclear power in Australia is unlikely to be economically viable.

Having tested the waters and found them apparently safe, the Prime Minister declared in February 2006 that he had an open mind on the subject of nuclear power for Australia, and that it was just a question of economics. 'I have no hang-ups at all about nuclear energy', he declared, before perhaps saying more than he should have about the political strategy behind the nuclear push: 'It presents the most exquisite dilemma to the Greens, too'. Contradicting himself, he went on to declare:

> You can't ignore market forces and I think it makes a lot more sense for us to put dollars into finding ways of burying

carbon and cleaning up fossil fuel, given that we are the largest coal exporter in the world; we have these vast supplies of natural gas.[33]

Howard decided to take the debate to the next stage, but he did so while overseas as he met with US, Canadian and Irish government officials. In May he said from Ottawa: 'I want a full-blooded debate in Australia about this issue, and I want all of the options on the table'.[34] Appealing to the environmental advantages, he said that a nuclear industry in Australia was 'inevitable', not forgetting to drive the wedge in a little further: 'It is cleaner and greener, and therefore some of the people who in the past have opposed it should support it'.[35] Writing in *The Australian Financial Review*, Julie Macken argued that this formed part of a larger global push. While in North America, Howard was briefed on George W. Bush's plans for a Global Nuclear Energy Partnership under which Australia would become 'a one-stop nuclear shop that leases enriched uranium to encourage greenhouse-friendly power generation'.[36]

The Labor Party's response to the nuclear debate was a knee-jerk one. Betting that public softening of attitudes was overstated, Kim Beazley, the leader of the Opposition at the time, declared: 'There will be no nuclear power in Australia under a Beazley Government'.[37] The public debate, which until then had been flat, was fired up by the intervention of the Australia Institute, which in May 2006 named the seven sites in Australia most likely to host the first nuclear power plant. The siting criteria included the need to be near a major water source, which meant the east coast of Australia, a major load centre, the main transmission lines of the national electricity grid and a port for the importation of the fuel rods. The most likely sites included the Sunshine Coast north of Brisbane, Port Stevens on the New South Wales central coast, south of Wollongong, Westernport Bay to the east of the Mornington Peninsula, the western side of

Port Phillip Bay and Portland in Victoria. On 24 May the story appeared on the front page of *The Sydney Morning Herald* and a media frenzy ensued. Mayors, business leaders and local residents expressed their opposition. Various local members – including the parliamentary secretary for the environment, Greg Hunt – went into damage control, assuring constituents that the Prime Minister's nuclear power plants would be kept out of their backyards. It was not quite the full-blooded debate that Howard had envisaged.

Howard was playing a risky game. It makes no sense to develop a nuclear industry in Australia: it's too expensive and would take too long for it to have an appreciable effect on emissions. The real objective of raising it as an option was to open the way for an expansion of uranium mining and an enrichment industry. At the same time, the Government knew that Labor's position on uranium mining was contradictory and that it would prefer not to talk about the issue. Labor's 'three mines' policy was an attempt to accommodate anxieties in the party and among the public, which are particularly strong among women voters. However, few issues provoke such passion as nuclear ones, and the attempt to wedge Labor always carried the risk of alarming key constituencies unnecessarily.

Howard had a plan and he pressed on with it. In June 2006 he announced the formation of a task force on nuclear energy. Its remit was 'to undertake an objective, scientific and comprehensive review into uranium mining, processing and the contribution of nuclear energy in Australia in the longer term'. In his announcement Howard noted that a 'growing number of environmentalists now recognise that nuclear energy has several other advantages over fossil-fuel electricity generation, including significantly lower levels of air pollution and greenhouse emissions'.[38] Howard had already referred to Tim Flannery's apparent support for nuclear power in Australia.[39] The task force was chaired by Dr Ziggy Switkowski (trained as a nuclear scientist) and included Professors George Dracoulis and

Warwick McKibbin, both of the Australian National University. The fact that the group contained only supporters of nuclear power engendered some cynicism. In July, Howard said in a speech:

> With close to 40 per cent of the world's known low-cost uranium deposits, for Australia to bury its head in the sand on nuclear energy is akin to Saudi Arabia turning its back on global oil developments ... All sources of energy have advantages and disadvantages. The real question is whether Australia should fully consider its interests and responsibilities in the global nuclear energy debate or whether it succumbs to a dogma of denial.[40]

The Prime Minister who refused to take the nation into global efforts to cut emissions now had the chutzpah to appeal to our 'responsibilities in the global nuclear energy debate'.

The draft report of the nuclear task force was released in November.[41] Given the composition of the task force, the report was a reasonably balanced one and contained nothing that was not known or not predictable. It said that nuclear power would be between 20 and 50 per cent more costly to produce than electricity from coal or natural gas. It was optimistic about the timing, suggesting that nuclear electricity could be delivered to the grid in ten years, though a wait of 15 years was more likely. This contrasted with the finding of the Stern Review in the UK, which, as we shall see, stressed that in order to forestall the worst consequences of climate change we must begin cutting emissions within a decade.

The most striking part of the report was the presentation of a scenario in which 25 reactors would generate over a third of the nation's electricity by 2050. Unsurprisingly, the spectre of 25 nuclear power plants generated headlines. There was no escaping the question of where they would be sited. *The Sydney Morning Herald* carried the headline: 'The big issue: location, location, location'.[42] *The Herald-Sun* led with: 'Nuclear reactors would be built close to Melbourne and the Latrobe Valley under a blueprint for Australia's energy future',[43] and its stablemate in Sydney, *The Daily Telegraph*, had a double-page spread with a map showing possible sites for the reactors in New South Wales.[44] Labor immediately went on the offensive over the siting, and the task force chairman, Switkowski, said, 'We haven't wrestled with the issue of where individual reactors would be located'.[45]

When asked if he thought the proposal for 25 nuclear plants was realistic, the Prime Minister began dousing the fires, suggesting that it was just speculation and putting off the nuclear question for the future.[46]

MOUNTING PRESSURE

As public anxiety and media awareness about climate change grew in 2006, new interests were drawn into the campaign. Philanthropic organisations such as the Myer Foundation, which for years had ignored the issue as too hard, suddenly wanted to get involved. The Climate Institute (Australia) was established and its rural campaign struck a chord across farming communities that had been brought to their knees by the never-ending drought.[47] As a result, organisations such as the NSW Farmers Federation, which had for years harboured sceptics who saw climate change as another attempt by greenies to place limits on farmers' rights, began demanding real action by the Federal Government.

The Climate Institute also brought together almost all of the main religious groups in Australia to endorse a statement on the need for action.[48] As well as the established Christian denominations, it included Muslims, Buddhists, Sikhs and Jews. Politically, the most interesting inclusions were evangelical Christians and the influential Australian Christian Lobby. Climate change was explicitly framed as a moral issue. Anglican Archbishop George Browning declared that 'wilfully causing environmental degradation is a sin'; the Australian Evangelical Alliance said that taking action immediately was a matter of justice; and the Australian Christian Lobby wrote that 'Christians have a moral duty to be stewards of the creation' and said it would be looking carefully at the positions taken on climate change by the political parties in the run-up to the federal election expected at the end of 2007. This must have been worrying for the Government.

There was more. An international poll by the BBC in July 2006 found that among citizens of 19 countries, Australians were more worried than any others about climate change.[49] Asked how concerned they were about energy use causing environmental problems including climate change, 69 per cent of Australians said they were very concerned, with another 25 per cent somewhat concerned. Britons, Brazilians, Canadians and Italians all had at least 60 per cent very concerned, compared with 53 per cent of Americans and only 20 per cent of Russians.

No-one took any notice of the BBC poll. In contrast, a survey conducted in late June and early July 2006 and published by the Lowy Institute in October seems to have had a profound impact on Government thinking. Global warming had emerged as the big sleeper issue of national affairs.[50] Sixty-eight per cent of Australians now rated global warming as a 'critical threat' over the next decade, only just behind terrorism and rogue nuclear powers. For years the public had been worried about climate change but appeared to look for reasons not to accept the full implications of the scientists' warnings. By 2006 the message seemed finally to have sunk in. The director of the Lowy Institute, Allan Gyngell, noted that global warming had entered the mainstream. 'It's no longer just an issue for Greens and people dressed up in koala suits', he said, seemingly unaware that Australians had been expressing their concern in surveys for years. With a global agreement a decade old and a large market in emission credits already in operation, this indicated just how out of touch the foreign policy establishment in Australia had been.

The Government's own polling was showing that the public accepted the idea of Kyoto – four out of five wanted Australia to sign up – so the Government shifted from its stance of trashing the agreement at every opportunity and began to talk as if it accepted Kyoto but wanted to build on it. Following the lead of Ian Campbell, it conjured up the term 'new Kyoto', although there was no scheme, no grouping and no process associated with this term. In fact, it was no more than an incantation, one that left the rest of the world bemused. Labor's environment spokesperson Anthony Albanese had never heard of the 'new Kyoto' proposal, so he googled it. The only reference he could find was to the New Kyoto Hotel, of which, he gleefully told parliament, a reviewer had written: 'The worst aspect of the room was that the window didn't open and there is no way to cool the room down or get some fresh air'.[51] The rest of the world's climate representatives, gathered at Nairobi in November 2006 to discuss the Kyoto Protocol, were in the dark too and dismissed it as just another red herring from Australia.

Pressure on the Government over its obstinate stance was building. A Newspoll report commissioned by Greenpeace showed strong community demand for tougher action, with 80 per cent of those surveyed supporting the imposition of levies on greenhouse polluters. More frightening science emerged, with one study predicting that half of the world's major forests would be lost if global temperatures rose by 3°C by the end of the century.[52] Climate scientists now reluctantly conceded that temperatures would almost unavoidably rise by 2°C, causing severe impacts, but hoped that the catastrophic effects of a 3°C rise could be averted if the world were to move rapidly to start cutting emissions. The business community was losing its inhibitions about criticising Howard's inaction, with even the Business Council of Australia, and especially its new president, Michael Chaney, increasing pressure on the Government to get serious. A few months earlier, the Australian Business Roundtable on Climate, a powerful alliance of corporate executives brought together by the Australian Conservation Foundation, had produced a report arguing that Australia's emissions could be slashed while the economy continued to grow.

The publication of the Stern Review on 30 October 2006 attracted extensive media coverage around the world. Sir Nicholas Stern, a former chief economist at the World Bank, had been commissioned by UK chancellor Gordon Brown to write a comprehensive report on the economics of climate change. Stern's unwritten brief was to engage with the arguments used by the US and Australian governments to refuse to join international efforts to reduce emissions. The analysis of Stern and his team concluded that the economic costs of doing nothing, that is, the damage to economic activity of climate change, were likely to exceed the costs of deep cuts by an order of magnitude. In this way, he turned the economic argument against change on its head.

US and Australian defiance can in part be explained by the fact that any agreement to cut greenhouse gases would impinge on economic interests. Conservative administrations have been strong advocates of economic globalisation but have resisted efforts to pursue global responses to cross-border environmental impacts. As the Stern Review observed, global warming is the worst instance of 'market failure' yet seen. It pointedly referred to the problem of free-riding: implying that if global warming is the biggest externality we have ever faced, then the refusal by the United States and Australia to ratify the Kyoto Protocol represents the worst instance of free-riding we have ever seen.

By this point the Howard Government had been spooked by the

extraordinary turn of events and the clamour for it to take the issue more seriously. The nuclear ploy had clearly failed, and AP6 appeared dead in the water. Something new was needed. On 13 November, the Prime Minister told a Business Council of Australia dinner that he would establish a 'task group' to examine a carbon trading scheme. The group was stacked with executives from the greenhouse polluting industries. Oddly, its terms of reference did not ask it to develop a proposal for a domestic emissions trading scheme but a global one. Since, after years of negotiation, the Kyoto Protocol had already established a global trading system – which includes developing countries by way of their capacity to generate emissions credits through the Clean Development Mechanism – the group appeared to have been given the job of creating an alternative to Kyoto, as though a handful of fossil-fuel executives in Australia could do in six months what it had taken the nations of the world a decade to achieve. Howard was floundering.

The reaction in Nairobi to the Prime Minister's announcement of a task force on emissions trading betrayed the fact that policy was being made on the run in Canberra. The Australian delegation in Nairobi was as surprised as everyone else when the news came through. More significantly, the chief lobbyists for the fossil-fuel industries, John Hannagan and John Daley, refused to believe the news when told by a smiling Anthony Albanese and 'turned white' when the news sank in that a major greenhouse decision to consider carbon trading had been made without their knowledge.

Signs of confusion in the Government were reinforced by the performance of Treasurer Peter Costello on ABC's *Lateline* program on 16 November. In a rambling and incoherent performance, he claimed to favour a new approach to climate change that incorporated a global emissions trading system. When asked if that would include China and India, he said that it would be unfair to ask China to cut its emissions when rich countries have grown rich by burning fossil fuels and are largely responsible for the problem. This, of course, is precisely why industrialised countries had agreed at the Kyoto conference that only they should have mandatory emissions reductions in the first commitment period, and precisely why the Howard Government had rejected the protocol. As he responded to pointed questioning by Tony Jones, the Treasurer seemed to understand the absurdity of his Government's position and began to giggle.

It was instructive to observe the reaction of the anti-greenhouse commentators to the chaotic events of October and November, and the apparent shift in the Government's position. Writing in *The Australian*, Christopher Pearson, who without any relevant qualifications had been setting himself up as an expert on climate science, desperately sought an explanation for the Prime Minister's 'bitterly disappointing' surrender.[53] He should have taken a 'bolder stand right from the start of the debate', opined Pearson. He should have sacked weak-minded environment ministers and 'promoted more of the informed debate we have seen in the pages of *The Australian* from the likes of Bjorn Lomborg and Bob Carter'. Howard's capitulation, he argued, had its origin in the flawed character of the Australian people, who have become 'credulous' in place of the admirable 'scepticism' of their grandparents. *They* would not have been steamrolled by the scaremongers of the press and the 'claque' of scientists. Pearson is the same man who has regularly congratulated the Australian public for their good sense in electing John Howard.

Pearson lauded the intervention of Nigel Lawson, who had launched a blistering attack on the Stern Review and the 'new religion of eco-fundamentalism'. In order to give him credibility, Pearson wrote that Lawson was Margaret Thatcher's 'most notable chancellor of the exchequer', but he seemed to have forgotten that Thatcher was the very first world leader to raise the alarm about climate change.

The glaring contradictions in his argument reveal the true

motivation of sceptics such as Pearson and fellow right-wing commentators Piers Akerman, Andrew Bolt and Alan Wood. Pearson endorsed Lawson's view that environmentalism 'is profoundly hostile to capitalism and the market economy'. This is the nub of the matter. The sceptics' logic is as follows:

- environmentalists are the enemies of capitalism;
- what they advocate must be contrary to the interests of capitalism;
- those who provide the evidence that supports environmentalists' views (climate scientists and policy makers) are also enemies of capitalism;
- accepting the evidence of global warming means giving in to anti-capitalists; therefore
- we must not accept the science of climate change and will seek out any shred of evidence that appears to contradict it.

The bitter disappointment of Pearson, who has been close to the Prime Minister, was evidence of the panic in Government ranks. The Government had not primed its supporters for the shift. But it also indicated that the Government had no clear idea where it was going on the issue and was casting around for immediate relief from political pain.

Unlike his colleagues, *The Australian*'s editor-at-large Paul Kelly has rarely had difficulty jettisoning positions when it suited him. Prone to pontificating, he did not bother attempting to defend his previous repudiation of climate change and, without missing a beat, moved straight on to defend the new position of the Government, seeking to provide it with a strategic rationale. Howard's volte-face was not a capitulation, declared Pope Paul, but a signal that Howard the master politician was going on the offensive and stealing the march on his opponents. 'Those who say Howard is a belated convert miss the point.'⁵⁴ In Kelly's world every defeat for Howard is turned into a cunning strategic victory, if only ordinary observers had his unparalleled political insight.

Yet Kelly's analysis was so full of elementary errors about the history and nature of international climate policy that a critique is scarcely possible. For example, his faith in the ability of his Prime Minister to turn AP6, the would-be rival to Kyoto, into a new global emissions trading system (something that even the Prime Minster had not at that point advocated) seemed to fly in the face of two facts. First, the Bush Administration was cool on AP6 and Congress had twice refused to fund it. Secondly, and more importantly, four of the six members of AP6 were already participating in a global emissions trading system, not least India and China, and were receiving billions of dollars in investments under the Kyoto Protocol's Clean Development Mechanism.† But Kelly did not allow facts to nobble good spin, and he glibly dismissed a decade of the most difficult global negotiations ever to take place as merely a 'symbol', implying that the powers at Howard's fingertips would be enough to induce the world to repudiate that effort and embrace the 'new Kyoto', an empty slogan dreamt up by the leader of a small nation in the South Pacific. Kelly even reproduced the Government's absurd claims that the protocol would have no effect on greenhouse gas emissions. If, instead of listening to the Prime Minister's spin-masters, Kelly deigned to attend one of the annual conferences of the parties, open himself to the views of the thousands of experts and negotiators from around the world, and attend one or two seminars, he would cringe in embarrassment at the distortions he has reproduced in column after column, year after year.

---

† In February 2007, the UNFCCC secretariat announced the registration of the 500th project under the CDM, noting another 950 in the pipeline. The lion's share of investment is going to Brazil, China, Mexico and, especially, India (UNFCCC secretariat, *Kyoto Protocol clean development mechanism passes 500 registered project milestone*, press release, 12 February 2007).

During December, fires raged across Victoria, Tasmania and South Australia. When asked about a report linking climate change to increases in the frequency and severity of bushfires the Prime Minister dismissed the claims as 'esoteric'.[55] The link between global warming and increased bushfire risk had been established in a series of reports and papers by the CSIRO, the Bushfire Cooperative Research Centre, the Council of Australian Government's national inquiry into bushfire mitigation and the just-released Commonwealth State of the Environment Report. CSIRO projections indicated that climate change will increase the frequency of very high and extreme fire danger days by between 4 and 25 per cent by 2020 and between 15 and 70 per cent by 2050 across south-east Australia.[56] Howard's dogged refusal to accept the advice of scientists is consistent with his scepticism about global warming. The fires brought home the message of the Stern Review that the costs of climate change are likely to dwarf the costs of cutting emissions.

As 2006 drew to a hot and smoky close, the Prime Minister obtained fleeting relief from an unexpected source. Some of Australia's leading climate scientists were surprised to read in *The Australian* or hear on the television news of a new study from the CSIRO that claimed 'the drought gripping southeast Australia is due to natural variations in climate rather than the greenhouse effects'.[57] It quoted the conclusions of a CSIRO researcher, Barrie Hunt, who had modelled the pattern of drought in Australia over the last 10,000 years. The story featured prominently on all television stations, and other newspapers followed up on *The Australian*'s scoop the next day. The story countered claims, including those by South Australian Premier Mike Rann, that the crippling drought was being exacerbated by climate change.

Those inclined to climate scepticism seized on the story. On 2GB Jason Morrison, filling in for Alan Jones, used it to talk about how 'lies get propagated about this climate change phenomena'. In his blog, *The Daily Telegraph*'s Piers Akerman said the report undermined the 'hysteria' among 'hothouse extremists'.[58] Naturally, *The Australian* devoted its editorial to it, claiming that 'CSIRO research fellow Barrie Hunt has done everyone a service by blowing the whistle on the pessimistic hand-wringing that accompanies too much of the discussion on climate change', going on to denounce the 'political blather' and 'unproven theories' of those concerned about climate change.[59]

All of this was welcome news for the Government, and it was reasonable to suspect that it may have orchestrated the intervention. So how did the story get into the press? Hunt himself had approached *The Australian* with the story based on his work over the previous few years. Hunt had worked for many years as a climate scientist at the CSIRO's Division of Atmospheric Research in Aspendale. He retired in 2004 but kept his office in an honorary position at the organisation. As journalists probed him over the next few days, it emerged that Hunt's scientific views were more complex than the early news stories had suggested. As the former head of the CSIRO's climate modelling program, he is not a greenhouse sceptic and conceded that there are clear signs that global warming is making the drought worse by drying out soils.[60] He claimed that his statements about the drought were motivated by a concern that when the rains return, people might stop worrying about climate change. Other eminent scientists have said that there is no way that the conclusion that the current drought was due to natural variability alone could be drawn from Hunt's modelling work.[61] The global warming trend and changes in rainfall patterns are superimposed on natural variability.

Hunt's backtracking could not undo the impact of the initial stories, which must have sown doubt in the minds of many Australians, adding once again to the tempting conclusion that the scientists can't make up their minds so there is no need to worry. Barrie

Hunt himself was well known at the CSIRO for his 'very conservative' political views, ones he often expressed. He is a long-standing and active member of the Liberal Party. As well as serving as the contact for the Frankston East branch, he has sought preselection on more than one occasion.[62]

PRAYING FOR RAIN

As 2007 got under way, it was apparent that the Prime Minister had thought long and hard over the holiday break. It was now obvious even to his loyal supporters that his approach to climate change had failed. It was no longer going to be possible to fob off the public with dismissive words and token policies. The people were worried, even alarmed. The Labor Party was getting traction in the electorate on global warming, and the weather would be one of the top three issues for the election, expected in October or November. It wouldn't be the first election in which the weather played a part. In Britain it used to be said that as an election approached, the Conservative Party would pray for rain. In a voluntary voting system, if it rained on election day the Tories could get more cars on the road to take older voters to the polling booths. The Coalition Government was also praying for drought-breaking rain; in early 2007 relief was in sight as the Bureau of Meteorology began forecasting an autumn end to El Niño.

The first step for Howard was to sideline his embattled environment minister and promote the ambitious and capable Malcolm Turnbull into the job. Tellingly, the job was renamed: while Labor had a shadow minister for climate change, the recently appointed Peter Garrett, Turnbull would be the Minister for Environment and Water Resources. The new name signalled the strategy Howard had worked out to seize control of the climate change debate, or at least to loosen Labor's grip on it. He was aware that for most Australians the most worrying manifestation of climate change was the drought. The two had become joined in the popular understanding, both in the bush where the paddocks were parched and in the cities where water restrictions had turned lawns brown. So Howard decided to turn the climate problem into a water problem. Water was something he could be seen to be doing something about. Give the people their water and the votes will flow.

In response to a carefully orchestrated media campaign, on the day after Australia Day the newspapers all led with the Government's announcement of a bold $10 billion plan; in exchange for the states ceding control of the Murray-Darling Basin to the Commonwealth, the latter would spend that sum on water projects, including $3 billion on buying back over-allocated water licences from irrigators.[63] The Prime Minister would be seen to be doing something about the most obvious manifestation of climate change in this country. Already he had begun executing his new strategy; whenever asked about climate change and his Government's indefensible record, he would immediately begin talking about water. In his speech to the Press Club, he began by grudgingly acknowledging the prospect of long-term climate change, but immediately went on to declare: 'I regard myself as a climate change realist', which seemed to be the first step away from being a denialist but a long way from accepting the seriousness of the crisis. He then went straight into the drought and what he could do about it with no further mention of its causes.[64]

The new minister for water followed the same script. Asked about climate change, he immediately segued into water. For those hoping that the change of minister represented a change in the Government's position, Turnbull's first interview on ABC's *7.30 Report* was dispiriting. He wheeled out all of the rationalisations that Campbell had worn out and, like the Government as a whole, ended up with just one – blame China.[65] Unlike his predecessor, however, who was willing to deploy a stream of words to cover up his Government's neglect, Turnbull is a man of action, not content to be a window-dresser and dissembler, the traditional role of environment minister. Howard has always dictated greenhouse policy, just as he determines all major policy, and it will be diverting to see whether Turnbull is willing to do what all other ministers have done in order to succeed under Howard. One test will be whether Turnbull receives the same treatment as Campbell when it comes to setting policy. Campbell was never invited to the vital consultations between the Government and the fossil-fuel lobby, where the real agenda was agreed. Will Turnbull, too, be left outside the door awaiting his instructions?

Howard's new strategy of turning the climate change problem into a water problem was a response to a political imperative rather than a real-world one. Of course, spending $10 billion on water projects is not going to make one iota of difference to climate change; as an attempt to manage one manifestation of the climate crisis, it is a diversion from the real problem – greenhouse gas emissions. The strategy can do no more than ameliorate, for a few years, one aspect of a crisis that worsens by the year. Howard seems to believe that his obligation is to help Australians adapt to climate change rather than play a part in solving it.

A few days after the Prime Minister reshuffled his ministry, on 26 January 2007, he announced that Tim Flannery was Australian of the Year. For Howard, the award was both an opportunity and a threat. By embracing the choice of Flannery as Australian of the Year, Howard could signal that he really did care about global warming. What Flannery thought about having his popularity used to boost the environmental credentials of the prime minister who turned Australia into an international climate change pariah was another question entirely.

15. SABOTAGING THE FUTURE?

UNDERMINING KYOTO

Enough is now clear for us to lay out the Howard Government's overall strategy. There is no question that the Government decided early on that it would not be party to any international agreement to cut greenhouse gas emissions. It has also resolutely refused to introduce policies aimed at cutting emissions in Australia. The polluting industries have been protected, and so their emissions have increased. Other than those designed as window-dressing, the policy solutions the Government has proposed – geosequestration and nuclear power – will not have any significant effect for 15 to 20 years.

Yet this is not the whole of the story. Taken together, the evidence suggests that the Howard Government has not been content to refuse to take effective measures and to quarantine Australia from global attempts to do the same. Rather, it has actively set out to sabotage the Kyoto Protocol.[1] If this is indeed the goal, it explains the history of counter-moves and illogical and contradictory arguments for not signing up that the Government has used for many years now. Consider the facts as we know them.

The Government's economic modelling indicates that at least half of the economic costs to Australia of Kyoto would arise not from domestic action to cut emissions but from the actions of other parties as they attempt to meet their obligations. Australian coal exports would be in the front-line. This helps to explain why the

Government has repeatedly said it will not ratify because of the importance of protecting our energy exports, and especially our export coal industry, even though our energy exports would be wholly unaffected by policies to cut emissions in Australia. It is measures by other countries to cut their emissions that would damage our export coal industry.

If the objective is not only to avoid cutting Australia's emissions but also to stop other countries cutting theirs, the Government's repeated reference to protecting our energy exports makes sense. This argument dovetails with the Government's firm belief, repeated over and over, that Australia's future prosperity and strength as a nation depend on one factor above all others: our ability to increase energy exports to Asia. This was the central argument of both the 1997 white paper on trade and foreign affairs and the 2004 white paper on energy, which stated that 'developing Australia's abundant low-cost energy resources is a key to our future prosperity'. It was reinforced by the Prime Minister's claim in July 2006 that he wanted Australia to become an 'energy superpower'.[2]

Though formalised in the energy white paper, the idea that energy exports to Asia hold the key to Australia's future, and nothing should stand in their way, was at the centre of Howard Government thinking from 1996. In that year the position was put forcefully to a conference in Washington organised by the Competitive Enterprise Institute by Australian diplomat Paul O'Sullivan (now head of ASIO). Reminding the audience that Australia is the world's largest coal exporter, he said: 'Australia's trade outlook is also more and more defined by our growing economic ties with developing countries, particularly in the Asia-Pacific. Almost half (around 43 per cent) of our trade is with non-OECD countries and over 60 per cent of our exports go to Asia'.[3]

The special place of coal exports in the Government's world view was also revealed in the wholly disproportionate attacks on Greens leader Bob Brown when he suggested in February 2007 that coal should be phased out in a few years. In an extraordinary tirade, *The Australian*'s environment reporter Matthew Warren linked Brown's off-the-cuff remark to 'skyrocketing' energy prices, the death of trade-exposed industry, the drying up of investment, the loss of 'at least 200,000 jobs', 'devastated' communities, a collapsing Australian dollar, 'dramatic' import price rises, 'enormous' pressure on inflation, 'brutal application of monetary policy' and an economy sinking into 'recession or worse'. Then the real problems would arise – 'wages would be slashed', 'spiralling interest rates' would drive home-owners to 'foreclosures and fire sales', which would flow through to the share market, with Australians seeing their superannuation savings 'slashed'. 'Without question', Warren added, 'an economic hit on a scale that goes beyond living memory'. He finished by accusing the Greens leader of 'reckless hyperbole'.[4]

The strategy of sabotaging Kyoto is wholly consistent with the development of AP6. Despite the protestations of other parties, it is indisputable that the Howard Government wants the partnership to evolve into an alternative to Kyoto, a fact which environment minister Ian Campbell let slip and which is the subtext of everything the Prime Minister says on the subject. The emphasis on highly speculative and expensive 'clean coal' and carbon capture and storage can be understood as an attempt to protect the coal industry through a new international agreement. More generally, the Government's 'technological approach' is above all a delaying tactic; it wants to buy time so that coal-friendly answers can be found to the greenhouse problem. Australia's energy exports would then continue to have a secure future in a lower emissions world. This explains why a Government that insists that its first concern is with the economic costs of Kyoto at the very same time promotes the most expensive abatement solutions.

The hypothesis that the Howard Government has been actively

working to destroy the Kyoto Protocol also helps to explain a paradox that has bothered many commentators, including the European negotiators. If Australia has repudiated the Kyoto Protocol, why is it so desperate to continue to participate in Kyoto negotiations? Why does it want to shape a treaty that it has rejected outright? A plausible answer is that as long as the Howard Government has a seat at the table, it can continue to spoil and make progress more difficult.

The sabotage hypothesis also clarifies the peculiar role of China in the Howard Government's strategy. It is an argument that applies also to India, but with lesser force. Australia is the world's largest coal exporter. Although China is not at present a major market for Australian coal, the huge growth in its energy demand provides an irresistible lure, particularly as 80 per cent of our exports of steaming coal currently go to countries that have emission reduction obligations under the Kyoto Protocol (Japan, the European Union and, in due course, South Korea).[5] The Howard Government's claim that Kyoto is flawed because China does not have emission reduction obligations is deeply cynical because the last thing it wants is for China to have any such obligations. The Kyoto Protocol, which China has ratified and from which it is benefiting in the form of large flows of foreign capital under the Clean Development Mechanism, provides the best chance of getting China to commit to future emission reductions, which would choke off a lucrative future market for Australian coal.

While Australia vigorously attacks the protocol because it 'exempts' countries such as China and India from obligations to cut emissions, in fact the prospect of China taking on such obligations in the second or subsequent commitment periods is the Government's worst nightmare. Instead of sitting back and watching a huge future market evaporate, the Howard Government has developed a strategy to cultivate that market by sustaining the growth of coal-fired power in China. It believes that the best way to do this is to urge on the world an alternative to Kyoto that incorporates China, asks only for voluntary measures and places all emphasis on a 'technological approach' focused on 'clean coal' solutions. On becoming environment minister Malcolm Turnbull had learned the script. In his first media appearances he stressed that selling 'clean coal' technology to China 'may well be the most important thing Australia does in terms of greenhouse'.[6]

The Howard Government has the incentive to sabotage Kyoto, but does it have the power? Barry Naughten argues persuasively that the key to this is encouraging the United States to stay out of the protocol. Here Australia has played a significant role, providing cover for the Bush Administration's opposition. As we saw, Al Gore, among others, has suggested that without the backing of Australia, the United States would find it much more difficult to refuse to participate in global efforts. As long as the United States refuses to take on commitments, China (and other developing countries) will not contemplate doing so. On the other hand, if Australia were to become an enthusiastic participant in the Kyoto agreement, it is more likely that the United States would join up, and that would clear the way for China and other developing countries to consider adopting mandatory targets in the second or subsequent commitment periods of the protocol. In fact, the United States and Australia have reversed this sequence, arguing that they will not ratify unless China accepts a target, thereby knowingly creating a stalemate.

The Howard Government knows that Australians have for years been anxious about global warming and have wanted their Government to play a constructive role at home and abroad. Secrecy as to its real intentions has therefore been essential, and it has covered over its real agenda with a veneer of confected concern, a barrage of words and a handful of voluntary programs, the net effect of which on our greenhouse gas emissions has been virtually zero.

AN UNREAL DEBATE

It is painful to be a citizen of a nation that could behave in such an immoral way, but the evidence suggests that the Australian Government has deliberately harmed the only real prospect the world has of heading off the catastrophes that climate change is expected to visit on the Earth.

In preparing this book, I have read volumes of public and private correspondence between members of the greenhouse mafia and between them and the Government. It is truly striking that not once in their commentary have any of them expressed concern about climate change. There is nothing about the fate of poor people in developing countries, no sign of regard for those who may be displaced from island homes, no mention of the potentially devastating effect on the environment in Australia – the Barrier Reef, Kakadu, mountain ecosystems – or of the plight of farmers driven from their land by never-ending drought. Nothing. Only one issue preoccupies them: how to protect the profits of the fossil fuel–based industries. How do these people square what they are doing with their conscience? Do they ever worry about the appalling consequences if they are wrong?

At best, the debate in Australia has been about slowing the growth of our greenhouse gas emissions. Despite the apparent political difficulty of achieving even small reductions in emissions, there is an air of unreality about the debate so far, because we know from the scientists that we will need to cut global greenhouse gas emissions by at least 60 per cent and more likely 90 per cent by 2050 in order to prevent a dire situation becoming catastrophic. These deep cuts mean transforming the energy system that powers our everyday life. Like the climate system, energy systems have great inertia. It takes decades to replace or transform transport systems, electric power plants, buildings and the layout of cities; if we are to achieve deep cuts and avoid the worst effects of climate change, early planning and action are essential.

Although the Howard Government has attempted to persuade us that only new technology can save us from climate change, a number of studies in Australia and abroad have shown that we can achieve deep cuts with currently available energy technologies. To do this, we would need to invest heavily in energy efficiency, sharply reducing our demand for electricity and fuel for transport, and develop renewable energy industries – wind, solar and biomass – to replace fossil fuels. There are many social and environmental benefits from such an energy revolution – reducing urban air pollution, creating more and better jobs – such that solving climate change could also lead to an industrial renaissance. But there will be costs, too, and not just financial ones. We will need to get used to large areas being turned over to wind and solar power plants and more of our agricultural land being devoted to fuel crops.

We cannot wait for a radical technological breakthrough, such as nuclear fusion, cheap photovoltaics or solar photolysis of water, that will deliver abundant, low-cost and zero-emissions energy. Nor is the answer dreamt of by the Government and the coal industry – the development of cheap and effective technologies for carbon sequestration – available to us yet. If we were to put aside the other dangers, a partial solution would be to accept the development of several large nuclear power plants in Australia.

A more reliable, if less palatable, solution would be for Australians to accept policies that involve sweeping changes to their lifestyles, including restrictions on house design and use of household appliances, giving up some of our personal mobility including air travel, promoting telecommuting, shifting to higher density living, and changing our diets to consume less meat.

We have already lost a decade or more due to inaction. The greenhouse debate in Australia (as elsewhere) has fallen victim to short-term thinking. The electoral cycle, the immediate cost of structural adjustment and the impact on prices have dominated public discussion and the Australian Government's thinking. While the cynical view may be that governments never plan beyond the next election, this has not always been the case. In recent decades federal governments have shown themselves willing to embark on far-reaching reforms that result in major structural adjustment of the economy. Trade liberalisation, the floating of the exchange rate and the introduction of competition policy are the best examples of this. In each case, governments were willing to override the objections of vocal sectional interests in order to pursue what they believed was in the long-term interest of the nation. This ought to be grounds for optimism, yet the Howard Government has displayed a blind obstinacy on climate change, partly driven by a loathing of 'environmentalism' and partly by a desire to protect the export coal industry.

The story told in this book is one of greedy corporations and craven politicians; but it is also one of public disengagement. For a decade the anti-greenhouse forces have been able to delay effective action to cut emissions because of their ability to persuade citizens that they need not worry – the science is uncertain, the problem is exaggerated, someone else is to blame, we can do little ourselves and (at the same time) we have the problem under control. Until recently, although always uneasy, the public has been willing to go along with these arguments even in the face of mounting evidence from climate scientists about the disasters that could befall us if we fail to act.

In short, the Howard Government has been able to hoodwink the community with impunity because many Australians have preferred to believe the lies. The UK writer and activist George Marshall has argued that often, when confronted by potential catastrophes, intellectual knowledge is not enough, no matter how compelling it may be.[7] In order to act, people need to be shocked into a state of heightened awareness. There is an old German adage that helps explain why so many Jews stayed in that country in the 1930s: 'Things whose existence is not morally possible cannot exist'.[8] The consequences of failing to cut greenhouse gas emissions, as forecast by the climate scientists, are scarcely imaginable.

The history of climate change politics challenges the Enlightenment's faith in the power of science and reason. Humans are capable of a state of simultaneous 'knowing and not-knowing', a state typified by the position on climate change of the US and Australian governments. The Howard Government formally accepts the science of climate change and has officially endorsed the reports of the UN, CSIRO and so on – reports which, for any disinterested reader, make terrifying reading. Yet, judged by its actions, it believes that the threat of climate change is insignificant compared with the threat to Australia of other countries cutting their greenhouse gas emissions. Its approach may be summed up as, 'Do nothing at home and work hard to prevent others taking action'. Thus its policy apathy at home contrasts with its feverish activity abroad.

While the Australian public has indicated in surveys that it wants governments to take action, the Howard Government has found endless excuses for inaction, and until 2006 the public was lulled into a belief that somehow things would be all right. As both citizens and private consumers, the intensity of our concern has been strangely muted. George Marshall asks why the general public has been so quiescent on the issue of climate change.[9] Pointing to the various psychological processes that reinforce denial, he suggests that when our grandchildren ask us why we did nothing about climate change even though we understood what would happen, we will either deny knowledge ('I didn't know'), deny our agency ('I didn't do it'), deny our personal power ('I couldn't do anything') or blame others ('The corporations and George W. Bush did it'). These are the same excuses the fossil-fuel lobbyists and the Government ministers will use when history and their grandchildren ask why they refused to act.

It is now clear that most Australians are no longer in a state of denial, that they are facing up to the truth about global warming and what it means for life in this country and around the world. In these circumstances it will become more and more difficult for the Government and the fossil-fuel lobby to repeat their lies, distortions and spin with impunity. The same can be said for citizens in all of the nations of the industrialised world, and increasingly those of the developing world.

A decade has been lost, and we will pay dearly for it; but the next decade will see the beginning of the transformation of the world into one resolved to protect the Earth for future generations.

NOTES

CHAPTER 1
1. Guy Pearse, 'The business response to climate change: case studies of Australian interest groups', doctoral thesis, Australian National University, July 2005. I was one of the supervisors of the doctoral thesis.
2. ibid., p. 343, n. 671.
3. By 'the industry department' I mean that department with primary responsibility for minerals and energy. From 1987 to 1998 it was the Department of Primary Industries and Energy; from 1998 to 2002 it was the Department of Industry, Science and Resources; and from 2003 to 2006 it was the Department of Industry, Technology and Resources.
4. Pearse, p. 341.
5. ibid., p. 346, n. 675.
6. Earlier wins included: deleting almost all mention of greenhouse from the 1988 Energy Policy Statement; having crucial caveats inserted into the 1990 Interim Planning Target; and keeping discussion of a carbon tax out of the 1992 National Greenhouse Response Strategy.
7. Pearse, p. 349.
8. Personal communication, August 2001.
9. Pearse, p. 347, n. 675.
10. The environment department also leaked profusely in the early years of the Howard Government. The difference was that those leaks were plugged, while contact between public servants and the fossil-fuel lobby was tacitly endorsed.
11. Pearse, p. 318, n. 631.
12. ibid., p. 342, n. 667.
13. ibid., p. 352.
14. *The Sydney Morning Herald* website, 14 February 2006, <http://www.smh.com.au>.
15. Andrew Bolt, 'Sneaky green mafia', *Herald Sun*, 17 February 2006.
16. Sam Walsh, 'Notes of LETAG meeting with Prime Minister Howard, May 6 2004', 9 May 2004.
17. 'Industry communication on greenhouse policy in the PM's Energy Statement, DRAFT 2/6/04', sent by Lyall Howard, 4 June 2004, (Lyall.Howard@riotinto.com).
18. 'The greenhouse mafia', *Four Corners*, television program, ABC TV, 13 February 2006.
19. See Clive Hamilton & Sarah Maddison, *Silencing dissent*, Allen & Unwin, St Leonards, 2007.
20. Rosslyn Beeby, '"Climate of fear" in solar research', *The Canberra Times*, 30 May 2006.

21 Anne Davies, 'California dreaming, to stop a nightmare', *The Sydney Morning Herald*, 26–28 January 2007.

Chapter 2
1. Barrie Pittock, *Climate change: turning up the heat*, CSIRO Publishing, Collingwood, 2005, p. 79.
2. See, for example, ibid., pp. 8, 43.
3. Robert Watson, presentation at the Sixth Conference of the Parties to the UNFCCC, The Hague, 13 November 2000.
4. IPCC, *Climate change 2001: the scientific basis*, IPCC Third Assessment Report (Working Group I) summary for policy makers, IPCC, Geneva, 2001.
5. IPCC, *Climate change 2007: the physical science basis*, IPCC Fourth Assessment Report (Working Group I), IPCC, Geneva, 2007.
6. Pittock, *Climate change*, pp. 17–18.
7. ibid., p. 18.
8. K. Hennessy et al., 'Climate change in New South Wales, part 2: projected changes in climate extremes', consultancy report for the New South Wales Greenhouse Office, CSIRO, Aspendale, 2004, p. 6, table S2.
9. For an overview see Tony McMichael, Rosalie Woodruff and Simon Hales, 'Climate change and human health: Present and future risks', *Lancet*, vol. 367, 11 March 2006.
10. Australia, House of Representatives, *Statements by members*, 31 August 2000, p. 19912.
11. V. Brajer, 'Recent evidence of the distribution of air effects', *Contemporary Policy Issues*, vol. 10, no. 2, 1992, pp. 63–71; J. Donovan, 'Lead in children: report on the national survey of lead in children', Australian Institute of Health and Welfare, Canberra, 1996.

Chapter 3
1. UNFCCC, 'National greenhouse gas inventory data for the period 1990–2004 and status of reporting', UNFCCC, Bonn, 19 October 2006, <http://unfccc.int/resource/docs/2006/sbi/eng/26.pdf>. See also Hal Turton, *Greenhouse gas emissions in industrialized countries: where does Australia stand?*, Discussion Paper No. 66, The Australia Institute, Canberra, June 2004.
2. Turton, *Greenhouse gas emissions in industrialized countries*. One or two Middle Eastern oil producers, such as Kuwait and Oman, may have higher levels.
3. See Clive Hamilton, *Running from the storm*, University of New South Wales Press, Sydney, 2001, pp. 16–19.
4. Turton, *Greenhouse gas emissions in industrialized countries*, p. 14. Data in the next paragraph are from the same source.
5. Hugh Saddler, Frank Muller & Clara Cuevas, *Competitiveness and carbon pricing: border adjustments for greenhouse policies*, Discussion Paper No. 86, The Australia Institute, Canberra, April 2006.
6. ibid., p. ix.
7. Australian Coal Association, *Australian black coal exports*, ACA, Canberra, <http://www.australiancoal.com.au/exports.htm#Expsum>; Market Information and Analysis Section, DFAT, *Composition of trade Australia 2005–06*,

DFAT, Canberra, November 2006, <http://www.dfat.gov.au/publications/stats-pubs/cot_fy2006_analysis.pdf>.
8 International Energy Agency, Energy Policies of IEA Countries: Australia 1997 Review International Energy Agency/Organisation for Economic Co-operation and Development, Paris 1997.
9 L. Schipper et al., *Energy use in Australia in an international perspective*, OECD/IEA, Paris, 2001.
10 See especially a paper prepared for COAG: Graham Armstrong and Saturn Corporate Resources, *Preliminary assessment of demand-side energy efficiency improvement potential and costs*, National Framework for Energy Efficiency, Melbourne, November 2003.

CHAPTER 4
1 Office of National Assessments, *Fossil fuels and the greenhouse effect*, Economic Assessments Agency, Canberra, 1981.
2 R.J. Hawke, media release, 6 April 1989.
3 Ross Dunn, 'Keep nuclear option open: PM', *Australian Financial Review*, 7 April 1989.
4 The UNFCCC had been negotiated in 1991 and early 1992 but was opened for signature at Rio. After signature, countries must ratify any international treaty before it comes into force. In many countries ratification involves formal approval by the national parliament. In Australia, it is merely an executive act, although treaties must be tabled in parliament before ratification. Australia ratified the UNFCCC in December 1992 and the convention entered into force in March 1994, after the required 50 countries had ratified it.
5 COAG, *National Greenhouse Response Strategy*, AGPS, Canberra, 1992, p. 16.
6 *Greenpeace Australia v. Redbank Power Company Pty Ltd and Singleton Council*, (1995) 86 LGERA 143.
7 COAG, *National greenhouse response strategy*, Canberra, 1992, p. 13.
8 George Wilkenfeld, Clive Hamilton & Hugh Saddler, *Australia's greenhouse strategy: can the future be rescued?* Discussion Paper No. 3, The Australia Institute, Canberra, February 1995, p. 1.
9 Personal communication with a public servant in the department at the time.
10 Letter from Tony Beck, executive director of the AIGN, to Prime Minister Howard, 22 December 1997.
11 George Wilkenfeld and Associates & Economic and Energy Analysis, *Evaluating the greenhouse challenge: issues and options*, Greenhouse Challenge Office, Canberra, 1996.
12 It was done 'inadvertently' in the sense that he appeared unaware of the contents of the report and their significance.
13 In March 2005 the Government revamped the scheme with the launch of Greenhouse Challenge Plus. Reflecting the increased political pressure on the Howard Government, one of the more significant changes was the shift to mandatory membership. From 1 July 2006 Australian businesses that claim $3 million or more in fuel excise credits were required to sign up.
14 The following information in this section draws heavily on Richard Denniss, *Taxing concern? The performance of the Green Power Scheme in Australia*, Discussion Paper No. 31, The Australia Institute, Canberra, 2000.

15 ibid.
16 ibid.
17 NSW Department of Energy, Utilities and Sustainability (DEUS), *National greenpower accreditation program: quarterly status report*, no. 2, NSW DEUS, Sydney, April–June 2006.

Chapter 5

1 *The Canberra Times*, 25 May 1997.
2 *The Australian*, 24 July 1997, p. 3.
3 Alexander Downer, *Australia and climate change*, address to the Global Emissions Agreements and Australian Business Conference, Melbourne, 7 July 1997.
4 *PM*, radio program, ABC Radio, Sydney, 23 July 1997. See also reports on 24 July 1997 in: *The Sydney Morning Herald*, p. 10; *The Australian*, p. 3; *Australian Financial Review*, p. 5; and *The Age*, p. 7.
5 *The Sydney Morning Herald*, 25 June 1997, p. 4.
6 *Australian Financial Review*, 30 June 1997, p. 1.
7 *The Courier Mail*, 25 March 1998, p. 2.
8 Warwick Parer, *Coal utilisation and environmental management*, address to the UNDP senior executive seminar, Sydney, 17 February 1997.
9 See, for example, *The Age*, 30 July 1997, p. 8.
10 *The Sydney Morning Herald*, 14 March 1997, p. 8.
11 *Australian Financial Review*, 12 March 1998, p. 1.
12 ibid., p. 4
13 *The Canberra Times*, 23 March 1998, p. 1.
14 On 8 March 2000, *Australian Financial Review* reported that Parer 'has been appointed a director of several coal companies at the centre of conflict of interest allegations that plunged the Howard Government into crisis'.
15 The statement was initiated by Professors Peter Dixon, Tor Hundloe and John Quiggin, and myself. It and the names of the 131 signatories were read into the Senate Hansard by Senator Lees on 1 October 1997.
16 For a comprehensive critique see Clive Hamilton & John Quiggin, *Economic analysis of greenhouse policy: a layperson's guide to the perils of economic modelling*, Discussion Paper No. 10, The Australia Institute, Canberra, December 1997.
17 This is explained in more detail in Clive Hamilton & John Quiggin, *Economic analysis of greenhouse policy*.
18 Australia, Senate, *Questions on Notice*, no. 565, 15 May 1997, p. 3518.
19 Documents supplied to the ACF after a freedom of information request.
20 The quotations are from the report by the Commonwealth Ombudsman, *Report of the investigation into ABARE's external funding of climate change economic modelling*, Canberra, February 1998, pp. 9, 10.
21 Meeting documents of the AAC Executive Committee, Sydney, 28 October 1997.
22 Documents supplied to the ACF after a freedom of information request. Letter from Alan Powell to Brian Fisher, 16 July 1997.
23 Commonwealth Ombudsman, *Report of the investigation into ABARE's external funding*.

24 Personal communication from one who overheard the remark.
25 I discussed these issues in detail with three former DFAT officers intimately familiar with the climate change campaign.
26 DFAT, *Climate change: Australia's approach*, briefing paper, DFAT, Canberra, July 1997, p. 27.
27 Personal communication to a colleague from another department who repeated it to me.
28 Tech Central Station, *Alan Oxley*, TCS Daily, 2006, <http://www.tcsdaily.com/Authors.aspx?id=232>.
29 Michael Duffy, 'A cold, hard look at a hot topic', *The Sydney Morning Herald*, 9 April 2005.
30 See ExxonSecrets <www.exxonsecrets.org> and Source Watch, 'Australian APEC Study Centre', <http://www.sourcewatch.org/index.php?title=Australian_APEC_Study_Centre>.
31 Quoted in ACF media briefing, August 1997.
32 *The Age*, 30 July 1997.
33 See for example *The Canberra Times*, 25 June 1997, p. 1; *Australian Financial Review*, 23 September 1997, p. 4. On the basis of a one-page survey of companies and state government departments, DFAT's estimate of 90,000 lost jobs was arrived at by estimating the number of jobs expected from planned major investments over a five-year period ($68 billions' worth). 'If all of these projects were to proceed it is estimated that around 90,000 long-term jobs could be created. As a result of relative increases in cost pressures [due to greenhouse abatement measures] ... there would be possible reassessment of the viability of some of these projects' (DFAT quoted in *Australian Financial Review*, 29 September 1997, p. 3). Documents obtained by the ACF under freedom of information laws showed that some state governments distanced themselves from the employment claims. Nevertheless, on this feeble basis the Prime Minister, whipped into a frenzy by the BCA and the AIGN, claimed that 90,000 jobs would be lost if uniform targets were adopted.
34 *Australian Financial Review*, 14 November 1997.
35 John Howard, *Safeguarding the future: Australia's response to climate change*, statement to parliament, Canberra, 20 November 1997.
36 *The Sydney Morning Herald*, 26 November 1997, p. 1.
37 *Australian Financial Review*, 10 October 1997, p. 11.

CHAPTER 6

1 ABARE, *Developing countries the losers in emission abatement policies*, media release, 28 November 1997.
2 *The Australian*, 4 December 1997, p. 6.
3 Estimates of the extent of land-clearing in the crucial 1990 base year have varied widely from one inventory to the next. For a detailed analysis see Andrew Macintosh, *The national greenhouse accounts and landclearing: do the numbers stack up?*, Research Paper No. 38, The Australia Institute, Canberra, January 2007.
4 Recounted to me by former senior public servants.
5 According to an industry insider: 'Hannagan and Bushnell were ... funded by the energy intensive interests to go to the COPs at Geneva and Kyoto as

observers; this sort of co-ordinated activity would have started in about 1995 … and was funded by AIGN'. (Pearse, p. 326, n. 643.)
6 *The Australian*, 12 December 1997, pp. 1, 5.
7 *The Sydney Morning Herald*, 1 December 1997, p. 1.
8 *Australian Financial Review*, 13 December 1997, p. 31.
9 *The Sydney Morning Herald*, 12 December 1997, p. 1.
10 *The Sydney Morning Herald*, 19 December 1997, p. 10. In a letter to me in mid-1998, Bjerregaard wrote in reference to Australia's target: 'I hope I can rely on the [Australia] Institute and its partners to press for full and immediate action to implement the targets – *however unsatisfactory* – agreed in Kyoto' (emphasis added).
11 *The Sydney Morning Herald*, 18 December 1997, p. 13.
12 *The Sydney Morning Herald*, 19 December 1997, p. 10.
13 These observers include current and former DFAT officials who must remain anonymous.
14 Sebastian Oberthur & Hermann Ott, *The Kyoto Protocol: international climate policy for the 21st century*, Springer-Verlag, Berlin, 1999, pp. 52, 71.
15 ibid., pp. 137–38.
16 Document seen by me.
17 The inclusion of 'forestry' with land-use change complicates the picture a little but does not change the conclusions.

Chapter 7

1 *The Washington Post*, 3 March 1998.
2 ibid.
3 ibid.
4 *Dow Jones Newswires*, 21 April 1998.
5 *Dow Jones Newswires*, 21 May 1998.
6 *The Washington Post*, 30 April 1998.
7 *The Washington Post*, 3 March 1998.
8 *The Washington Post*, 18 October 2000.
9 Robert Manning & Susan Tillou, *The Los Angeles Times*, 1 March 1998.
10 See Leonie Haimson in *Grist Magazine*, 11 January 2001.
11 *The Los Angeles Times*, 1 March 1998.
12 *The Daily Telegraph*, 17 January 1998.
13 See <www.pewclimate.org>.
14 See Clean Edge Inc., *Clean Energy Trends*, Clean Edge, 2006, <http://www.cleanedge.com/reports-trends2006.php>.
15 See Leonie Haimson in *Grist Magazine*, 11 January 2001.
16 *The Financial Times*, 29 January 2001.
17 IPCC, *Land-use, land-use change, and forestry*, Cambridge University Press, Cambridge, 2000.
18 *The Financial Times*, 28 November 2000.
19 *Grist Magazine*, 22 June 2001.
20 *Business Week*, 9 April 2001. One American energy boss who has been a long-time observer of the debate said that he was amazed by 'the manner and tone of how all this has been handled: it was inexperienced and immature' (*The Economist*, 5 April 2001).

21 ABCNEWS, 17 April 2001, <http://www.ABCNEWS.com>.
22 Quoted by Leonie Haimson, *Grist Magazine*, 11 January 2001.
23 National Research Council, *Climate change science: an analysis of some key questions*, National Academy Press, Washington, 2001. See also K. Seelye & A. Revkin, 'Panel tells Bush global warming is getting worse', *The New York Times*, 7 June 2001.
24 BBC News Online, BBC News, London, 9 October 2001, <http://news.bbc.co.uk>.
25 BBC News Online, *Sinking islands urged to accept migrants*, BBC News, 13 November 2001, <http://news.bbc.co.uk/1/hi/world/asia-pacific/1653472.stm>. Kiribati had agreed to an Australian request to house 500 asylum seekers on the remote atoll of Kanton.
26 *Australian Financial Review*, 20–21 September 1997.
27 The comments were reported in *The Weekend Australian*, 8–9 June 1996, p. 8.
28 Farhana Yamin quoted by Gordon Hamilton, *Vancouver Sun*, 18 March 1998.

Chapter 8

1 Letter from Tony Beck to the Prime Minister, 22 December 1997.
2 Email dated 28 October 1998. There were only five recipients: Barry Jones (APPEA), Keith Orchison (ESAA), Dick Wells (Minerals Council), David Buckingham (BCA) and the David Coutts (AAC).
3 Richard Baker, 'Climate change: how big energy won', *The Age*, 30 July 2005.
4 *Australian Financial Review*, 26 March 2001.
5 Bridson Cribb (of the Pulp and Paper Manufacturers Federation), faxed memo to Dick Wells (MCA), Barry Jones (APPEA) and David Coutts (AAC), 6 August 1998.
6 *The Canberra Times*, 24 September 1998, p. 1. In a letter to me dated 15 October 1998, a spokesman for the European Union wrote: 'Making ratification of the protocol dependant upon ratification by other parties could lead to a deadlock situation where everybody waits for the others to move first. Such a situation would have clearly detrimental effects on the environment and as such would be in contradiction with the objectives of the Convention'.
7 According to US Embassy sources the topic of the speech was not pre-planned.
8 David Coutts, executive director of the AAC, fax to Barry Meyer, 31 July 1998.
9 See Commonwealth Department of Industry, Science and Resources, *Australian Energy News*, vol. 18, December 2000.
10 Warwick J. McKibbin, 'Modeling results for the Kyoto Protocol', report to the AGO, 15 March 2002, revised 5 April 2002.
11 McKibbin, ibid., tables A2.1, A2.2 and A2.3.
12 *Australian Financial Review*, 12 October 2000; *The Age*, 12 October 2000. *The Age*'s report was particularly uncritical, reproducing the most alarming numbers and containing no dissenting opinion.
13 *The Morning Bulletin* (Rockhampton), 27 October 2001.
14 See Clive Hamilton, Alan Pears & Paul Pollard, *Regional employment and greenhouse policies*, Discussion Paper No. 41, The Australia Institute, Canberra, October 2001.
15 Media release, 12 March 2002, <http://www.ecdel.org.au/pressandinformation/ClimateChange2.htm>.

16 Australian Greenhouse Office, *National emissions trading: establishing the boundaries*, Discussion Paper 1, Commonwealth of Australia, Canberra, 1999.
17 Tim Flannery, *The weather makers*, Text Publishing, Melbourne, 2005, p. 302.
18 Quoted by Matt Price, *The Weekend Australian*, 29–30 October 2005, p. 20.
19 Flannery, p. 306.
20 Michael Maniates, 'Individualization: plant a tree, buy a bike, save the world?' in Thomas Princen, Michael Maniates & Ken Conca (eds), *Confronting consumption*, MIT Press, Cambridge, 2002, p. 57.
21 Peter Wilmoth, 'Tapping into green mainstream', *The Age*, 14 January 2007.

Chapter 9

1 See *Australian Financial Review*, 18 January 2001, and Bill Nagle's evidence before the Joint Standing Committee on Treaties on 3 November 2000.
2 Australian Gas Association, *Assessment of greenhouse gas emissions from natural gas*, research paper no. 12, AGA, Canberra, May 2000.
3 Australian Gas Light Pty Ltd, *Submission to the Joint Standing Committee on Treaties inquiry into the Kyoto Protocol*, AGL, Canberra, 18 September 2000.
4 See, for example, its submission to the Joint Standing Committee on Treaties inquiry into the Kyoto Protocol, dated 15 September 2000, where it argues that the Kyoto Protocol is 'a coercive document which transcends national sovereignty'.
5 In a letter to Greenpeace, Westpac chief David Morgan wrote: 'The Kyoto agreement represents a key step in the international process and we agree with the aims and objectives of the Kyoto Protocol in reducing harmful emissions that contribute to global warming ... As a member of the Business Council of Australia, Westpac is also supporting the current review of the council's position on the Kyoto Protocol and Australian ratification'. Quoted in N. Wilson, 'Hot letter puts Kyoto on different page', *The Australian*, 18 February 2003.
6 Miranda McLachlan, 'Big business splits over greenhouse', *Australian Financial Review*, 22 November 2006.
7 The letters were later leaked to Sophie Black of *Crikey*. See her 'While the world burns, business leaders fiddle', *Crikey*, 30 October 2006.
8 Much of the information in this section is drawn from Hal Turton, *The aluminium smelting industry: structure, market power, subsidies and greenhouse gas emissions*, Discussion Paper No. 44, The Australia Institute, Canberra, January 2002. See also Hal Turton & Clive Hamilton, *Subsidies to the aluminium industry and climate change*, submission to the Senate Environment References Committee Inquiry into Australia's response to global warming, The Australia Institute, Canberra, November 1999.
9 For example in the AAC submission to the Joint Standing Committee on Treaties inquiry into the Kyoto Protocol, 15 September 2000.
10 Hal Turton, *The aluminium smelting industry*.
11 Alan Stockdale, *State withdraws from negotiations with Alcoa*, media release, Treasurer's Office, Canberra, 18 July 1995.
12 Sources are cited in Hal Turton, *The aluminium smelting industry*.
13 In 1994 the Goss Labor government in Queensland sold the Gladstone power station to Comalco for around 60 per cent of its book value, thereby providing

a long-term effective subsidy to the Boyne Island smelter (see Hal Turton, *The aluminium smelting industry*).
14 Hal Turton, *The aluminium smelting industry*.
15 'Any shift of production offshore would ... undoubtedly... increase global greenhouse gas emissions,' Foreign Minister Alexander Downer, media release, Canberra, 15 August 2002.
16 See Hal Turton, *The aluminium smelting industry*.
17 ibid.
18 Center for Clean Air Policy, <http://www.ccap.org/international/developing.htm>.
19 John Stanford, Allen Consulting, 'Greenhouse emissions trading', conference paper presented to Emissions Trading Conference, Melbourne, January 2000.
20 See, for example, *Environmental finance*, September 2001. The Tokyo Electric Power Company has invested in forest plantations in New South Wales and Tasmania with a view to generating carbon credits to offset its emissions at home.
21 John Quiggin, *More on Kyoto*, John Quiggin, 19 August 2002, <http://johnquiggin.com/index.php/archives/2002/08/19/more-on-kyoto/>.
22 'The ALS [Australian Libertarian Society] continues to challenge the global warming fear-mongers in their calls for drastic and costly government action.' (ALS, *Activities*, ALS, 2006, <http://australianlibertarian.wordpress.com/activities>.)
23 Piers Akerman, 'Red herrings line path to election', *The Daily Telegraph*, 19 November 2002.
24 PMSEIC, *From defence to attack: Australia's response to the greenhouse effect*, PMSEIC working group report, Canberra, 25 June 1999, p. 3.
25 *Lateline*, television program, ABC TV, 18 May 2005.
26 S. Peatling, 'Scientist quits top job to toil for miner', *The Sydney Morning Herald*, 17 May 2005.
27 Senate Employment, Workplace Relations and Education References Committee, *Inquiry into the office of the chief scientist*, Report 5, Commonwealth of Australia, August 2004.
28 Bob Brown, *Chief scientist must go – Brown*, media release, Canberra, 5 August 2004.

CHAPTER 10
1 See George Monbiot, *Heat: how to stop the planet burning*, Allen Lane, London, 2006, pp. 31 ff.
2 ibid., p. 32.
3 ibid., p. 39.
4 Quoted by Ross Gelbspan, *Boiling point*, Basic Books, New York, 2004, p. 41.
5 George Monbiot, 'Beware of the fossil fools', *The Guardian Weekly*, 6–12 May, 2004.
6 SourceWatch, *Tech Central Station*, SourceWatch, 2006, <http://www.sourcewatch.org/index.php?title=Tech_Central_Station#Related_SourceWatch_Resources>.
7 The DCI Group sold Tech Central Station to the editor of the site, Nick Schultz, at the end of 2006. Schultz was the political editor for FOXNews.com, the website of the Fox News Channel. See <http://www.prwatch.org/node/5371>.

8. SourceWatch, *DCI Group*, SourceWatch, 2006, <http://www.sourcewatch.org/index.php?title=DCI_Group>.
9. The Competitive Enterprise Institute, *About CEI*, CEI, 2006, <http://www.cei.org/pages/about.cfm>.
10. R. Brunet, 'It just ain't so, say these reputable scientists', *Alberta Report*, vol. 24 (48), 1997, pp. 20–21.
11. Exxonsecrets, *Factsheet: Competitive Enterprise Institute, CEI*, Exxonsecrets, 2006, <http://www.exxonsecrets.org/html/orgfactsheet.php?id=2>.
12. Mick Brown, 'Wiki's world', *The Weekend Australian Magazine*, 9–10 December 2006.
13. John Garnaut & Jane Counsel, 'Different shades of Hugh', *The Sydney Morning Herald*, Money & Business section, 17–18 August 2002, pp. 45, 49.
14. S. Mayne, 'Totally addicted to coal', *Crikey*, 10 February 2006.
15. Australian Electoral Commission, *Annual disclosure returns*, AEC, 2006, <http://fadar.aec.gov.au>.
16. Garnaut & Counsel.
17. ibid.
18. ibid.
19. Quoted in Garnaut and Counsel.
20. Quoted in Pearse, p. 342, n. 667.
21. Hugh Morgan, 'The dire implications of Coronation Hill', *IPA Review*, vol. 44, no. 4, Institute of Public Affairs, 1991.
22. 'I'm a great believer in promoting freedom of ideas and debate.' Hugh Morgan quoted in Elizabeth Knight, 'Morgan: why I shouldn't be sacked', *The Sydney Morning Herald*, 5 February 1994.
23. Hugh Morgan, 'Carbon blackmail doesn't lead to a greener future', *The Australian*, 10 June 2002.
24. A. Cornell, 'Why Ray Evans is always right', *Australian Financial Review*, 8 January 2005.
25. ibid.
26. SourceWatch, *Ray Evans*, SourceWatch, 2006, <http://www.sourcewatch.org/index.php?title=Ray_Evans>.
27. S. Mayne, 'Totally addicted to coal'.
28. See Peter Vincent, 'The debate hots up', *The Sydney Morning Herald*, Supplement, 30 October 2006.
29. Bob Burton & Sheldon Rampton, 'Thinking globally, acting vocally: the international conspiracy to overheat the earth', *PR Watch*, vol. 4, no. 4, 1997, <http://www.prwatch.org/prwissues/1997Q4/warming.html>.
30. Bob Burton, 'Wise guys down under: PR's economic-front moves on Australia', *PR Watch*, vol. 4, no. 4, <http://www.prwatch.org/prwissues/1997Q4/wise.html>.
31. Cooler Heads Coalition, *IPCC summary leaked; Australia turning against Kyoto; coal vote split on Bush, Gore*, 1 November 2000, <http://www.globalwarming.org/article.php?uid=229>.
32. Competitive Enterprise Institute, 2006, <www.cei.org>.
33. Cooler Heads Coalition, *Antarctic cooling down; the Antarctic ice sheet is growing; Hansen downgrades warming threat*, 23 January 2002, <http://www.globalwarming.org/article.php?uid=192 >. I was the critic.

34 Quoted by Peter Vincent, 'The debate hots up'.
35 A. Cornell, 'Why Ray Evans is always right'.
36 The quotations in this paragraph were recorded at the conference by Shane Rattenbury of Greenpeace.
37 ibid.
38 Lavoisier Group, *Submission to the Joint Standing Committee on Treaties inquiry into the Kyoto Protocol*, Lavoisier Group, Melbourne, 2000.
39 Evans wrote that the energetic officials of the AGO are 'committed to bringing Australia into the Kyoto tent, regardless of our national interest', *The Canberra Times*, 24 January 2002.
40 Quoted by Melissa Fyfe, 'Global warming – the sceptics', *The Age*, 27 November 2004.
41 Paul Pollard, report on Lavoisier Group Conference, Melbourne, 11 September 2001. The attendance list included Sir Arvi Parbo, Ian Castles, Harold Clough and one person who, on the lunch attendance list, identified as a vegan.
42 Bureau of Meteorology, *Climate change*, Commonwealth of Australia, 2007, <http://www.bom.gov.au/climate/change>.
43 Personal communication with me.
44 House of Representatives Standing Committee on Agriculture, Fisheries and Forestry, transcript of Melbourne hearings, 8 April 2003.
45 David Karoly, email to Mike Manton and John Zillman, 6 August 2003.
46 David Karoly has since said that his actual words were that he 'did not know the reason why, but that it might be due to political pressures'.
47 John Zillman, *Climate change: a natural hazard?*, speech at the book launch, Melbourne, 22 November 2004, <http://www.lavoisier.com.au/papers/articles/johnzillmanlaunch.htm>.

CHAPTER 11

1 See Union of Concerned Scientists, *Smoke, mirrors and hot air: how Exxon-Mobil uses big tobacco's tactics to manufacture uncertainty on climate change*, UCS, Cambridge, MA, January 2007.
2 Exxonsecrets, *Factsheet: Frontiers of Freedom Institute and Foundation, FoF*, Exxonsecrets, 2006, <http://www.exxonsecrets.org/html/orgfactsheet.php?id=35>.
3 William Kevin Burke, 'The wise use movement: right-wing anti-environmentalism', *Public Eye*, June 1993.
4 George Monbiot, 'The revolution has been televised', *The Guardian*, 18 December 1997; George Monbiot, 'Far left or far right?', *Prospect Magazine*, 1 November 1998.
5 Frank Furedi, *Politics of fear: beyond left and right*, Continuum International Publishing Group Ltd, 2005.
6 See, for instance, Frank Furedi, 'Meet the Malthusians manipulating the fear of terror', *Spiked*, 27 June 2006.
7 *Counterpoint*, radio program, ABC Radio, 21 February 2005.
8 Deborah Hope, 'Humanist with a stir in many camps', *The Australian*, 25 March 2006.
9 Editorial, 'A vision for the nation's future', *The Australian*, 21–22 October 2006.

10 Peter Saunders, 'Tofu terror added to list of reasons to be fearful', *The Sydney Morning Herald*, 17 January 2007.
11 Bjorn Lomborg, *The skeptical environmentalist*, Cambridge University Press, UK, 2001.
12 Gregg Easterbrook, 'Finally feeling the heat', *The New York Times*, 24 May 2006. Easterbrook still swallowed the Bush Administration's line that Kyoto was too cumbersome and was being ignored even by its signatories.
13 He was a keynote speaker at the Australian Institute of Energy national conference in Sydney in 2001.
14 C. Marriner, 'Minister likes Danish eco-sceptic', *The Sydney Morning Herald*, 3 October 2003.
15 *PM*, radio program, ABC Radio, Sydney, 11 August 2006.
16 Alan Wood, 'Chicken little greenies harm their own cause', *The Australian*, 5 March 2002.
17 The judgment was made by the official Danish Committee on Scientific Dishonesty. After a complaint by Lomborg, the Ministry of Science, Technology and Innovation ruled that the finding was invalid and asked the committee to revisit the case. In December 2003, after consideration the committee decided not to do so and closed the case. Danish Committees on Scientific Dishonesty, Danish Agency for Science and Technology Innovation, 2003, <http://forsk.dk/portal/page/pro4/FIST/FORSIDE/UDVALGENE_VIDENSKABELIG_UREDELIGHED/NYT_FRA_UVVU/PRESSEMEDDELELSER/FINAL_DECISION_LOMBORG>.
18 Bjorn Lomborg, *The Economist*, 2 August 2001.
19 Wikipedia, *Mark Steyn*, Wikipedia, 2006, <http://en.wikipedia.org/wiki/Mark_Steyn>.
20 *Counterpoint*, radio program, ABC Radio National, Sydney, 7 August, 2006.
21 Centre for Independent Studies, 2006, <http://www.cis.org.au/>.
22 Matt Price, 'Bizarre future in neo-con humour', *The Australian*, 19 August 2006.
23 Sian Powell in 'Strewth', *The Australian*, 26 October 2006, p. 12.
24 Andrei Illarionov, 'Protocol is just lots of hot air', *The Australian*, 9 December 2004.
25 Union of Concerned Scientists, *Smoke, mirrors and hot air*.
26 The following account draws on Kath Wilson, 'Activists: how to beat them at their own game', *Newsletter*, The Australia Institute, Canberra, September 2005.
27 Michael Duffy, 'Putting the heat on global warming', *The Daily Telegraph*, 25 December 2004.
28 Miranda Devine, 'A debate begging for more light', *The Sydney Morning Herald*, 2 March 2006.
29 Hal Turton & Clive Hamilton, *Population growth and greenhouse gas emissions: sources, trends and projections in Australia*, Discussion Paper No. 26, The Australia Institute, Canberra, December 1999; *The Age*, 4 November 1999.
30 Newspoll Market Research, *Greenhouse gas*, study prepared for Greenpeace Australia Pacific. Of the remaining 20 per cent, 10 per cent were opposed and 10 per cent said they did not know.
31 Peter Schwartz & Doug Randall, *An abrupt climate change scenario and its*

implications for United States national security, Global Business Network, California, October 2003.
32 This point is made by Bruno Latour, 'Why has critique run out of steam? From matters of fact to matters of concern', *Critical Inquiry*, vol. 30, no. 2, Winter 2004.
33 'The celebrity environmentalist', *FHM*, November 2006.
34 Clara Iaccarino, 'The hottest spots on the planet', *Sun-Herald*, 4 June 2006.
35 Heather Hansen & Kimberly Lisagor, 'Going, going ... go now', *Wish Magazine*, October 2006. None of the seven wonders was in Australia.

CHAPTER 12

1 *The Sydney Morning Herald*, 6 January 2006.
2 See Senator Ian Campbell, *Bald Hills wind farm project in the balance*, media release, 6 October 2004, <http://www.deh.gov.au/minister/env/2004/mr06oct204.html>.
3 See Senator Ian Campbell, *Bald Hills wind farm and cumulative impact study*, transcript of press conference, 5 April 2006, <http://www.deh.gov.au/minister/env/2006/tr05apr06.html>. Campbell gave his news conference in Perth, where there were no journalists with any knowledge of the situation.
4 Rob Hulls, Victorian Planning Minister, speaking on *The 7.30 Report*, television program, ABC TV, Sydney, 17 April 2006.
5 Andrew Macintosh, *Environment Protection and Biodiversity Conservation Act: an ongoing failure*, research paper, The Australia Institute, Canberra, July 2006.
6 *The Australian*, 24–25 June 2006 and again on 22 December 2006
7 Ewin Hannan, 'Lawsuit in wind as farm gets nod', *The Australian*, 22 December 2006.
8 *AM*, radio program, ABC Radio, Sydney, 29 June 2006; Ewin Hannan, 'Wind farms a total fraud: Nats minister', *The Australian*, 29 June 2006.
9 ABC Online, 'Greens to campaign against Bald Hills wind farm', 15 August 2006, <http://abc.net.au/news/newsitems/200608/s1715375.htm>.
10 Rick Wallace, 'People power to blow Lib over', *The Australian*, 25 September 2006.
11 *The Age*, 10 December 2005. 'Internationally, I would say we're in the top half-dozen nations in terms of being a constructive player and trying to find a way forward' (quoted by Matt Price, 'Why minister makes skeptics see green', *The Weekend Australian*, 29–30 October 2005).
12 Comments made in the recording of the *Insight* television program, SBS, Sydney, January 2006.
13 *The Australian*, 5 January 2006 p. 4.
14 Wikipedia, 'Mohammed Saeed al-Sahaf', Wikipedia, 2006, <http://en.wikipedia.org/wiki/Muhammed_Saeed_al-Sahaf>.
15 *The Weekend Australian*, 29–30 October 2005, p. 20.
16 Ian Campbell, *The World Today*, radio program, ABC Radio, Sydney, 29 November 2005.
17 Anne Davies, 'Appeal on green ruling likely', *The Sydney Morning Herald*, 29 November 2006.
18 *The Sydney Morning Herald*, 12 January 2006, p. 2.
19 ABC Radio, 19 November 2005.

20 *The World Today*, radio program, ABC Radio, Sydney, 8 December 2005.
21 Will Steffen, *Stronger evidence but new challenges: climate change science 2001–2005*, Australian Greenhouse Office, Department of the Environment and Heritage, Commonwealth of Australia, March 2006 (released in May 2006).
22 Senator Ian Campbell, *New report shows stronger evidence for climate change*, media release, Canberra, 23 May 2006.
23 Eloise Dortch, 'Homes at risk from rising sea', *The West Australian*, 26 May 2006.
24 Eloise Dortch, 'Rising sea level theories "flawed"', *The West Australian*, 30 May 2006.
25 Eloise Dortch, 'Science is debatable in the world of Senator Campbell', *The West Australian*, 31 May 2006.
26 John Howard, interview with Alan Jones, Radio 2GB, Sydney, 15 November 2006.
27 'They are increasing their generation capacity every five or six months by an amount that's equal to ours.' (*The 7.30 Report*, television program, ABC TV, Sydney, 8 February 2007.)
28 K. Fredriksen, *China's role in the world: is China a responsible stakeholder?*, Office of Policy and International Affairs, US Department of Energy, presentation before the US-China Economic and Security Review Commission, Washington, 4 August 2006.
29 It was brought to their attention by Catherine Fitzpatrick of Greenpeace.
30 James Button, 'Prickly China tales diplomatic swipe at Australia', *The Age*, 17 November 2006.
31 Senator Ian Campbell, media releases, Canberra, 12 and 16 November 2006.
32 The sources for this analysis are various close observers in Europe. See also Sophie Black, 'How Australia is privately begging to be let inside the Kyoto tent', *Crikey*, 17 November 2006.
33 Advice provided to me.
34 *The Sydney Morning Herald*, 5 January 2006.

Chapter 13
1 Gabrielle Walker, 'The tipping point of the iceberg: could climate change run away with itself?', *Nature*, no. 441, 15 June 2006, pp. 802–05.
2 The following account is based on Walker, ibid.
3 ibid., p. 804
4 For a brief summary see Barrie Pittock, 'Are scientists underestimating climate change?', *Eos*, no. 34, 22 August 2006.
5 Steve Connor, 'If we fail to act, we will end up with a different planet', *The Independent*, 1 January 2007.
6 Joint Science Academies' statement, *Global response to climate change*, June 2001.
7 Quoted by Odile Blanchard & James Perhaus, 'Does the Bush Administration's climate policy mean climate protection?' *Energy Policy*, vol. 32, no. 18, December 2004, p. 1993.
8 Delegation of the European Union to Australia and New Zealand, *Press and Information*, <http://www.ecdel.org.au/pressandinformation/Climate-

Change2.htm>.
9 Quoted in Climate Institute, *Top ten tipping points on climate change*, Climate Institute (Australia), Sydney, July 2006, p. 17.
10 ibid., p. 16.
11 United States Senate, Sense of the Senate Resolution 312, 109 Congress, Session 1, June 2005.
12 Editorial, 'Climate Shock', *The New York Times*, 27 June 2005.
13 Senator Olympia Snowe, *G8 leaders challenged to act on climate change*, media release, Washington, 30 June 2005.
14 Wendy Frew, 'The time bomb is ticking', *The Sydney Morning Herald*, 13 December 2005.
15 Prime Minister John Howard on 28 July 2005 at a press conference in Sydney to announce the pact.
16 Environment minister Senator Ian Campbell speaking to Nick McKenzie on *AM*, radio program, ABC Radio, Sydney, 7 October 2005.
17 Remarks reported by Reuters. See <http://www.planetark.com/dailynews-story.cfm?newsid=31833>.
18 Steve Lewis, 'Climate fund's $100 million kickstart', *The Australian*, 7–8 January 2006.
19 *The Australian*, 5 January, 2006.
20 A special correspondent, 'A climate of confusion', *The Australian*, 14 January, 2006.
21 Ian Plimer, 'Global warming a damp squib', *The Australian*, 5 January, 2006.
22 Amanda Griscom, 'Little pact or fiction? New Asia-Pacific climate pact is long on PR, short on substance', *Grist Magazine*, 4 August 2005.
23 ibid.
24 John Howard, transcript of *The Prime Minister The Hon John Howard MP, address to the Asia-Pacific Partnership on clean development and climate*, inaugural ministerial meeting, Sydney, 12 January 2006.
25 Brian Fisher et al., *Technological development and economic growth*, ABARE Research Report 06.1, ABARE, Canberra, January 2006, p. 34.

CHAPTER 14

1 *The Sunday Age*, 28 May 2006.
2 Al Gore on *Enough Rope*, television program, ABC TV, Sydney, 11 September 2006.
3 *AM*, radio program, ABC Radio, Sydney, 11 September 2006.
4 Tracy Ong, 'Minister supports Gore's science', *The Australian*, 18 September 2006.
5 *The 7.30 Report*, television program, ABC TV, Sydney, 11 September 2006.
6 *Four Corners*, television program, ABC TV, Sydney, 28 August 2006.
7 Review of *An inconvenient truth* on *At the movies*, television program, ABC TV, Sydney, 6 September 2006. Pomeranz was rebuked by viewers and posted a partial *mea culpa* on the program's website.
8 'It's not the end of the world', *The Australian*, 4 September 2006.
9 *The Australian*, 5 August 2006.
10 Quoted by Sourcewatch, <http://www.sourcewatch.org/index.php?title=Matthew_Warren>.

11 Gaby Hinsliff, 'The PM, the mogul and the secret agenda', *The Observer*, 23 July 2006.
12 Sophie Black in *Crikey*, 12 September 2006 and *The Observer*, Sunday 23 July 2006.
13 'Sexy Ruth backs going green', *The Sun*, 11 September 2006, <http://www.thesun.co.uk/article/0,,2006410004-2006420081,00.html>.
14 Anne Henderson, 'It's all a bit rich being hectored by celebrity hypocrites', *The Australian*, 13 September 2006.
15 Matthew Warren, 'Climate horror flick could have been even scarier', *The Australian*, 11 September 2006.
16 Matthew Warren, 'Too vital for guesses', *The Weekend Australian*, 2–3 September 2006.
17 Matthew Warren, 'Forest needed to offset greenhouse from footy', *The Australian*, 20 September 2006.
18 *The Economist*, 7 September 2006.
19 *Business Review Weekly*, 22 September 2006.
20 See Jeffrey Sachs, 'Fiddling while the planet burns', *Scientific American*, 14 September 2006.
21 James Murdoch, 'Business has to do more to tackle climate change', *The Guardian*, 25 September 2006.
22 David Adam, 'Royal Society tells Exxon: stop funding climate change denial', *The Guardian*, 20 September 2006.
23 Terry Macalister, 'Oil giant hits at "unfair" attack by scientists', *The Guardian*, 22 September 2006.
24 Clayton Sandell, 'Senators to Exxon: stop the denial', *ABC News*, United States, 27 October 2006, <http://abcnews.go.com/Technology/story?id=2612021>.
25 Union of Concerned Scientists, *Smoke, mirrors and hot air*.
26 'Exxon spends millions to cast doubt on warming', *The Independent*, 14 December 2006.
27 Alan Wood, 'Doomsayers should stick to the facts', *The Australian*, 28 January 2003. See also 'Debate on climate change far from over', *The Australian*, 19 July 2006 and 'Greenie alarmists miss target', *The Australian*, 3 May 2006.
28 Alan Wood, 'Ignore the doomsday prophets', *The Australian*, 25 October 2006.
29 John Howard, press conference, Canberra, 6 June 2006.
30 Brendan Nelson, Dame Pattie Menzies Oration, Sydney, 18 April 2005.
31 S. Peatling, 'Nuclear power only natural, says Nelson', *The Sydney Morning Herald*, 11 August 2005.
32 Alexander Downer, 'Australia, the global environment and the economy', 2005 Sir Condor Laucke Oration, Barossa Valley, 1 September 2005.
33 Laura Tingle, 'PM backs nuclear solution', *Australian Financial Review*, 24 February 2006.
34 Denis Shanahan & Steve Lewis, 'PM urges unity on uranium', *The Australian*, 20 May 2006.
35 ibid.
36 Julie Macken, 'Out, then back: the big N-plan', *Australian Financial Review*, 7 June 2006.
37 Kim Beazley & Martin Ferguson, *Nuclear power in Australia*, media release – joint statement, Canberra, 23 May 2006.

38 John Howard, *Review of uranium mining processing and nuclear energy in Australia*, media release, Canberra, 6 June 2006.
39 *The 7.30 Report*, television program, ABC TV, Sydney, 6 June 2006.
40 John Howard, 'The nuclear debate we have to have', *The Sydney Morning Herald*, 18 July 2006.
41 Department of the Prime Minister and Cabinet, *Uranium mining, processing and nuclear energy – opportunities for Australia?*, draft report, Commonwealth of Australia, Canberra, 2006.
42 W. Frew, 'The big issue: location, location, location' *The Sydney Morning Herald*, 22 November 2006.
43 B. Packham & M. Harvey, 'Nuclear push powers ahead blueprint for reactors on city's doorstep', *Herald Sun*, 22 November 2006.
44 Malcolm Farr & Luke McIlveen, 'Don't force nuclear on us', *The Daily Telegraph*, 22 November 2006.
45 'Plan for 25 N-reactors in suburbia', *The Hobart Mercury*, 22 November 2006.
46 John Howard, doorstop interview, Ho Chi Minh City, 21 November 2006.
47 The Climate Institute was established with a $10 million donation from the Poola Foundation (Tom Kantor Fund). I was the initial chair of the board of the institute.
48 Climate Institute (Australia), *Common belief: Australia's faith communities on climate change*, CIA, December 2006, <http://www.climateinstitute.org.au/cia1/downloads/041206_common_belief.pdf>.
49 BBC World Service, '19 nation poll on energy', July 2006.
50 Peter Hartcher, 'Canberra, take note: climate change is what terrifies us', *The Sydney Morning Herald*, 3 October 2006.
51 Anthony Albanese in Australia, House of Representatives, *Matters of public importance*, 2006, p. 90.
52 Alok Jha, 'Forecast puts earth future under a cloud', *The Guardian*, 15 August 2006.
53 Christopher Pearson, 'Hotheads warned, cool it', *The Weekend Australian*, 18–19 November 2006.
54 Paul Kelly, 'New convert eyes a second coming', *The Australian*, 15 November 2006.
55 ABC Online, *PM dismisses climate change bushfire claims*, 18 December 2006.
56 K. Hennessy et al., *Climate change impacts on fire-weather in south-east Australia*, CSIRO, Canberra, 2005.
57 Asa Wahlquist, 'Greenhouse gases "not to blame"', *The Australian*, 28 December 2006; ABC Online, *Drought a result of natural causes, says researcher*, 28 December 2006.
58 <http://blogs.news.com.au/dailytelegraph/piersakerman/index.php/dailytelegraph/comments/drought_within_climate_norm>.
59 Editorial, 'Strange weather is situation normal', *The Australian*, 29 December 2006.
60 Liz Minchin, 'Rain won't end our problems: climate expert', *The Age*, 29 December 2006.
61 Personal communication with senior scientists.
62 <http://www.geocities.com/endeavour_uksa>. Hunt is also the president of the British Australian Community (formerly the United Kingdom Settlers'

Association), an organisation that claims to promote the 'common heritage' of British settlers who, it believes, face discrimination due to 'Anglophobia', described as 'a medical condition in which someone suffers an irrational fear of the English'.
63 For example, see Phillip Coorey, 'Howard rains $10 billion on rivers', *The Sydney Morning Herald*, 26–28 January 2007.
64 John Howard, 'The tyranny of the lowest common denominator must end', *The Sydney Morning Herald*, 26–28 January 2007.
65 *The 7.30 Report*, television program, ABC TV, Sydney, 25 January 2007. Turnbull said: 'China has got a lot more coal than we have, they mine a lot more coal, they burn a lot more coal; India is in a similar position. They are not going to abandon coal-fired power as an energy source. So if you're saying clean coal may not be a successful mission, well, maybe you'll be right, we'll have a look at it in ten years' time'.

Chapter 15

1 This has also been argued by Barry Naughten, 'Climate change: Howard holds a monkey wrench', *New Matilda*, 15 September 2006.
2 'As an efficient, reliable supplier of energy, Australia has a massive opportunity to increase its share of global energy trade. And with the right policies, we have the makings of an energy superpower.' (John Howard, address to the Committee for the Economic Development of Australia, Sydney, 18 July 2006.)
3 *Australia and climate change*, address by Paul O'Sullivan, deputy chief of mission, Australian Embassy, Washington, to the Competitive Enterprise Institute's conference 'The costs of Kyoto: the implications of climate change policy', Washington, 15 July 1997. <http://www.dfat.gov.au/environment/climate/0_sull.html>. O'Sullivan gave the same speech to the progressive US think-tank Resources for the Future.
4 Matthew Warren, 'Prepare for the crash', *The Australian*, 15 February 2007.
5 Barry Naughten, 'Climate change: Howard holds a monkey wrench'.
6 *The 7.30 Report*, television program, ABC TV, Sydney, 8 February 2007.
7 George Marshall, 'Denial and the psychology of climate apathy', *The Ecologist* (UK), November 2001. Marshall draws on the work of Stanley Cohen in *States of denial*.
8 Marshall cites Primo Levi as his source for this adage. See Primo Levi, 'Beyond Judgment', *New York Review of Books*, 17 December 1987.
9 Marshall, op. cit.

LIST OF ABBREVIATIONS

AAC	Australian Aluminium Council
ABARE	Australian Bureau of Agricultural and Resource Economics
ABC	Australian Broadcasting Corporation
ACA	Australian Coal Association
ACF	Australian Conservation Foundation
AGO	Australian Greenhouse Office
AIGN	Australian Industry Greenhouse Network
AOSIS	Alliance of Small Island States
AP6	Asia-Pacific Partnership on Clean Development and Climate
APPEA	Australian Petroleum Production and Exploration Association
BBC	British Broadcasting Corporation
BCA	Business Council of Australia
CO_2-e	carbon dioxide equivalent
COAG	Council of Australian Governments
CEI	Competitive Enterprise Institute
CEO	chief executive officer
CSIRO	Commonwealth Scientific and Industrial Research Organisation
DFAT	Department of Foreign Affairs and Trade
ESAA	Electricity Supply Association of Australia
G77	Group of 77 (developing countries)
GCP	Greenhouse Challenge Program
GDP	gross domestic product
GNP	gross national product
GWh	gigawatt hour
IPCC	Intergovernmental Panel on Climate Change

LETAG	Lower Emissions Technology Advisory Group
MCA	Minerals Council of Australia
MRET	Mandatory Renewable Energy Target
Mt	millions of tonnes
NGO	non-governmental organisation
NGRS	National Greenhouse Response Strategy
OECD	Organisation for Economic Co-operation and Development
ONA	Office of National Assessments
PR	public relations
PRIA	Public Relations Institute of Australia
RCP	Revolutionary Communist Party
TASSC	The Advancement of Sound Science Coalition
UN	United Nations
UNFCCC	United Nations Framework Convention on Climate Change
WMC	Western Mining Corporation

INDEX

Page numbers in *italics* indicate figures. An 'n' after a page number indicates a reference to a footnote.

A1B, 24n
A1F1, 24n
ABARE *see* Australian Bureau of Agricultural and Resource Economics
ABB, 85n
ABC, 147, 163, 194, 196, 212–13, 219
ABCNEWS (US), 89
Aborigines, Hugh Morgan on, 133
AC Nielsen-McNair, 69
ACA, 5, 9, 62
ACF, 6, 62, 160, 211
ACIL Consulting, 5
activism, workshop on how to defeat, 156–7
Ad Hoc Working Group on Future Commitments, 177
Adam Smith Club, 134
The Advancement of Sound Science Coalition, 128–9
Africa, 92
Against Nature (documentary), 147–9
The Age, 159, 163
AGL, 112–13
AGO, 36
agriculture, 34, 38, *38*, 40
AIGN, 5, 50, 96, 113, 116, 134, 136
air quality, 35
Akehurst, John, 8
Akerman, Piers, 123, 214, 217
Albanese, Anthony, 210, 212
Albright, Madeleine, 102
Alcan, 84
Alcoa, 10, 12, 85n, 115, 120, 121
Alcoa Foundation, 121n
Allen Consulting, 104, 105, 121

Allen, Nick, 138
Alliance of Small Island States, 31
alternative energy *see* renewable energy
alumina industry, 41
aluminium industry, 16, 40, 41, 115–21
aluminium smelters, 116–20, 134
AM (radio program), 194
amenity costs, 34
American Electric Power, 85
American Petroleum Institute, 131, 137
Anderson, John, 8
Andrews, Gwen, 71n, 99
Annan, Kofi, 176
Annex B countries, 77–8
Antarctica, 180
anthracites, 28
Anvil Hill, 172
AOSIS, 31
AP6, 187–92, 215, 223
APCO, 128
APPEA, 4, 102
Arctic sea ice, receding of, 180
Arnold, Ron, 148
Asia-Pacific Economic Cooperation Study Centre, 67
Asia-Pacific Partnership on Clean Development and Climate, 187–92, 215, 223
'assigned amounts', 77
astroturfing, 128, 156–7
At the Movies (television program), 195
Australia Institute, 159, 205
The Australian
    announces Asia-Pacific Partnership, 188

anti-climate science position of, 150, 152, 163–4, 189, 196–202
asks McKibbin about environmental refugees, 93n
calls Campbell's decision 'bizarre', 167
disappointed in Howard's softening position, 213
Hill writes to regarding Kyoto Protocol, 77
on the inclusion of the 'Australia clause', 74
publishes climate sceptics, 155, 156
publishes Morgan's praise for Howard, 135
seizes on CSIRO report as vindication, 217
Australian Aluminium Council, 4, 62–3, 98, 116, 120, 134, 136
Australian Bureau of Agricultural and Resource Economics
Aluminium Council on committee of, 116
analysis of AP6 by, 191, 192
economic modelling of, 57, 60–4
exaggerates cost of emission reductions, 68
executive director talks of evacuating islands, 94
issues paper on developing countries, 70–1
Lavoisier Group on conspiracy in, 140
loses credibility after economic modelling, 103
promotes Howard Government view of Kyoto Protocol, 55
receives award for modelling from PM, 96
Australian Business Council for Sustainable Energy, 114
Australian Business Roundtable on Climate, 211
Australian Christian Lobby, 209
Australian Coal Association, 5, 9, 62
Australian Coal Exporters, 59
Australian Conservation Foundation, 6, 62, 160, 211

Australian Democrats, 60, 166
Australian Eco-Generation Association, 114
Australian Evangelical Alliance, 209
Australian Financial Review, 60, 143, 145, 163, 205
Australian Football League, 200
Australian Gas Association, 112
Australian Greenhouse Office
administered by ministerial council, 97, 98
Andrews appointed to head, 99
attacked by fossil-fuel lobby, 98
commissions papers on emissions trading, 107
costing of compliance with Kyoto Protocol, 122
despised by Lavoisier Group, 142
Howard announces establishment of, 97
Howard never seeks advice of, 99
inventory of Australia's greenhouse gas emissions, 80
Morgan criticises, 140
submissions to Cabinet, 7
warns Parer of Kyoto implications, 102
Australian Industry Greenhouse Network, 5, 50, 96, 113, 116, 134, 136
Australian Institute of Petroleum, 5
Australian Labor Party, 93n, 125, 205, 206, 208
Australian of the Year, 220
Australian Petroleum Production and Exploration Association, 4, 102
Australian Public Service, 4
automobile industry, 84

bagasse, 108
Baillieu, Ted, 169
Bald Hills wind farm, 165–8
Baliunas, Sallie, 92n
Bangladesh, 23
Bank of America, 85n
Barrier Reef, 164
Batterham, Robin, 124–6
BBC Poll, 209
Beale, Roger, 73, 98

Beazley, Kim, 205
Beck, Tony, 96
Beckett, Margaret, 26
'belch tax', 105
Bennelong Society, 136
Berlin Mandate, 100
Bertram, Dean, 200
BHP, 51, 62, 114, 133
BHP-Billiton, 10, 133
bilharzia, 25
biofuels, 86
Bjerregaard, Ritt, 75–6
black coal, 28, 112
Blair, Tony, 58, 199
Boeing, 85
Bolt, Andrew, 9, 158, 198, 214
Bonn Conference (2001), 91
Boral, 10
Bourne, Greg, 114
Bracks, Steve, 170
Branson, Richard, 111
Brisbane Writers Festival, 150
British Petroleum, 82, 83, 85, 87n, 114
Brown, Bob, 125, 223
brown coal, 28, 112, 113
Brown, Gordon, 211
Browne, John, 83
Browning, George, 209
BSkyB, 199, 201
Buckingham, David, 8–9
Bungendore, New South Wales, 168
Bureau of Meteorology, 144–6, 178, 186–7
Burke, Brian, 178
Burton, Bob, 137
Bush Administration, 6–7, 23, 89–91, 129–31, 154, 182, 215
Bush, George W., 89–90, 182–3, 205
Bushfire Cooperative Research Centre, 216
bushfires (2006), 216
Bushnell, Noel, 67n, 113
Business Council of Australia
    Buckingham as executive director of, 8
    criticises Campbell, 167
    greenhouse mafia in, 6
    helps to fund ABARE's economic modelling, 62
    Howard announces carbon trading task force to, 212
    Morgan as president of, 134, 136
    pressures government on climate change, 211
    split by progressive stance of BHP, 114
Business Environmental Leadership Council (US), 120
Business Review Weekly, 201
Business Week, 89, 151

Calvert-Jones, John, 133
Campbell, Ian
    on *An Inconvenient Truth*, 195
    on the AP6 pact, 187
    attends Twelfth Conference of the Parties in Nairobi, 175–6
    blames China for emissions, 175–6
    claims figures vindicate Government policy, 80
    criticises Anvil Hill decision, 172
    dismisses emission targets, 172
    dismisses Kyoto Protocol, 170–1, 176
    epiphany in Tasmanian forest, 165
    on greenhouse emissions, 172–3
    on the 'new' Kyoto, 188, 210, 223
    not invited to vital consultations, 220
    puts spin on release of Steffen Report, 173–4
    replaced by Turnbull, 218
    suffers from Stockholm syndrome, 178
    on Tim Flannery, 109–10
    vetoes proposal for Bald Hills wind farm, 166–70
Canada
    as an exporter of energy, 41
    greenhouse gas emissions, 39, 40
    Kyoto Protocol targets for, 78
    melting of permafrost in, 181
    ratifies Kyoto Protocol, 94
    searches for loopholes in protocol, 87

use of hydroelectric power, 16
The Canberra Times, 102
carbon, 29
carbon capture and storage, 125, 126
carbon cycle, 72n
carbon dioxide, 27–8, 29, 37, 49
carbon emissions trading, 105–8, 114, 186, 212, 215
carbon sequestration, 227
carbon tax, 6, 50, 90, 105
Carboniferous period, 28
Carson, Rachel, 151
Carter, Bob, 20, 92n, 140n, 200
Carteret Islands, Papua New Guinea, 93n
Castles, Ian, 140
C.D. Kemp Lecture, 155
CDM, 106, 175, 212, 215, 224
cement industry, 16
Cement Industry Federation, 5
Centre for Independent Studies, 123, 150, 152, 155
Centre for the New Europe, 202
Chaney, Michael, 211
Channel 4 (UK), 147, 148
Charles River Associates, 137n
Cheney, Dick, 90
China
    blamed by Australia as emissions culprit, 175–6, 224
    Costello on, 212–13
    emissions cuts in, 175–6, 224–5
    greenhouse gas emissions, 39
    joins AP6, 187
    participates in CDM, 215
    signs Asia-Pacific Partnership, 190
Chrysler, 85
Church, John, 174
Cigna Corporation, 132
Clark, Sarah, 163
'clean coal', 59, 225
Clean Development Mechanism, 106, 175, 212, 215, 224
climate change
    aluminium industry opposes measures to combat, 115–21
    corporations begin to accept science of, 82–3
    denialists of, 20–2, 127–58
    economics of, 211
    inertia of, 25
    media failure to focus on, 161–4
    outline of sceptics' logic, 214
    power struggles over climate change policy, 98
    public apathy regarding, 228–30
    public opinion on, 158–61, 209–10
    relation to drought, 216–17
    role of climate science, 19–22
    'tipping points', 179–82
    weather changes in 2005, 179
    see also global warming; greenhouse gas emissions; Howard Government; science
Climate Institute (Australia), 208–9
climate science *see* science
Clinton Administration, 57, 89, 90, 154, 185
Clinton, Bill, 58, 82, 93
Clough, Harold, 139
coal
    black coal, 28, 112
    brown coal, 28, 112, 113
    Brown suggests phasing out of, 223
    'clean coal', 59, 225
    coal reserves, 30
    as a source of electricity, 16, 119–20
coal industry, 16, 28–9, 42, 48–9, 59, 221–2
    see also fossil-fuel lobbyists
Coastal Guardians, 168
Comalco, 124
combined-cycle gas turbines, 17
Committee for the Economic Development of Australia, 77
Commonwealth Department of Foreign Affairs and Trade, 55, 64–6, 155
Commonwealth Heads of Government Meeting (2002), 91–2
Commonwealth Ombudsman, 63
Commonwealth Scientific and Industrial Research Organisation, 13, 26–7, 144, 216–17
Commonwealth State of the Environment Report, 216

compensation, for victims of pollution, 32
competition policy, 228
Competitive Enterprise Institute, 131–2, 137, 138, 202, 222
Conference of the Parties
  Fourth Conference, 86
  Sixth Conference, 87–8, 91
  Eleventh Conference, 170, 185, 186, 188
  Twelfth Conference, 175–6
'congealed electricity', 117
Connaughton, James, 192
'contrarians', 20–2
Cook, Robin, 58
Cool Communities, 109
Cooler Heads Coalition, 132, 138
COPs *see* Conference of the Parties
Cormack Foundation, 133
Coronation Hill, 134–5
Corporate Europe Observatory, 202
Costello, Peter, 100, 133, 157, 170, 212–13
The costs of Kyoto ... (Adler), 137n
Council of Australian Government, 216
Countdown to Kyoto conference, 66–7, 113, 116, 138
Counterpoint (radio program), 150, 155
The Courier-Mail, 196
Court, Richard, 178
Coutts, David, 4, 102, 120
CRA, 51, 62
Cribb, Bridson, 97–8
Crichton, Michael, 157
Crikey, 198
crop yields, 92
CSIRO, 13, 26–7, 144, 216–17
Cusack, Barry, 134
Cutajar, Michael Zammit, 75

The Daily Telegraph, 151, 208, 217
Daimler-Benz, 83
Daley, John, 5, 212
Dargaville, Jackie, 169
Darwin, Charles, 130
Davos meeting (2000), 87

The day after tomorrow (film), 179
DCI Group, 131
Deakin University, wind power forum, 168–9
deforestation, 27
dengue fever, 25
Denmark, Western Australia, 168
Denniss, Richard, 53
developing countries
  aluminium smelters in, 119
  effects of climate change in, 25
  Howard Government claims regarding, 56, 100
  Howard Government spreads fear among, 70–1
  inspired to use Australia's example, 76
  in the Kyoto Protocol, 29–30
  principles of justice in treatment of, 31–2
Developing countries the losers in emission abatement politics, 70
Devine, Miranda, 152, 158
DFAT, 55, 64–6, 155
differentiation, 55–8, 79
Dingell, John, 67
disease, 25
Dobriansky, Paula, 91
Domenici, Pete, 185
Dow Chemical, 132, 157
Downer, Alexander, 57, 100–1, 155, 172, 204
Dracoulis, George, 206
Driscoll, Bob, 12
drought, 216–17, 219
Duffy, Michael, 150, 155, 157–8
Duke Energy, 85n
DuPont, 84, 85n
Durkin, Martin, 149

Earth Summit, 47
Easterbrook, Gregg, 151
EBCO Corp, 132
economic growth, 154
economic modelling, 60–4, 72, 96, 103, 121, 191, *192*, 221
economic rationalism, 109
The Economist, 151, 201

Edison Mission Energy, 10, 12
Egypt, 23
El Niño effect, 23, 162, 218
electricity, 16, 40, 112, 119–20
electricity industry, 17–18, 48, 107–8, 113
Electricity Supply Association of Australia, 5, 48, 62, 113
Ellis, Liz, 194
emissions *see* greenhouse gas emissions
emissions trading *see* carbon emissions trading
Energex, 10
Energy Australia, 51, 52
energy efficiency, 19, 43, 48
Energy Research and Development Corporation, 60
energy resources, export of, 222
Environment Australia, 98
Environment Protection and Biodiversity Conservation Act, 98
Environmental Manager, 163
environmental movement, 6, 160
Environmental Protection Agency (US), 128
Environmental Protection and Biodiversity Conservation Act, 166
environmental refugees, 13, 93, 94
ESAA, 5, 48, 62, 113
Estrada, Raoul, 71, 74, 75
ethics, 31–4
European Commission Delegation to Australia, 105–6
European Emission Trading System, 105
European Union
    calls for 15 per cent reduction in emissions, 65
    emission reductions under Kyoto Protocol, 37
    emissions trading system, 186
    greenhouse emissions in, 78–9
    oil companies in, 83–4
    ratifies Kyoto Protocol, 94
    reaction to American stance on Kyoto, 89
    unconvinced by Australia's submission, 177
Evans, Gareth, 5
Evans, Ray, 67n, 132, 133, 136–9, 146, 158
exchange rate, floating of, 228
exports, 41, 42, 65–6, 222
ExxonMobil
    criticised by James Murdoch, 201
    funds ABARE economic modelling, 62
    funds anti-green front organisations, 67, 129–30, 132, 138, 156, 158, 201–2
    refuses to change position on climate change, 83
    Royal Society writes to, 202
Eyles, John, 101

The failure of the Kyoto process (Kellow), 140
fairness, principles of, 31–4
Farquhar, Graham, 72n
fertilisers, 28
FHM, 164
The Financial Times, 87, 156
Finkel, Alan, 111
Fischer, Tim, 57, 68
Fisher, Brian, 62, 94, 137
Flannery, Tim, 109–10, 206, 220
Fleischer, Ari, 182
Florida, cyclones in, 35
Ford Motors, 83, 84, 85, 86, 137
Ford, William, 84
Fortune 500 survey, 87
fossil-fuel lobbyists
    AGA makes break with, 112
    alarmed at emissions trading plans, 107
    attack Australian Greenhouse Office, 98
    attempt to discredit science of climate change, 20
    back Greenhouse Challenge Program, 50
    create impression of scientific disagreement, 195–6
    fund ABARE economic modelling, 62–3

greenhouse mafia in, 3–12, 66, 116, 226
influence on Howard Government, 3, 6–12, 15, 18, 66
lack of social concern, 226
Parer reveals Cabinet decision to, 102
pleased with Kyoto outcome, 96
political power of, 43
send delegate to Kyoto, 74
shift blame to developing countries, 100
stage Countdown to Kyoto conference, 66–7
threaten supporters of Kyoto Protocol, 115
work to undermine Kyoto deal, 99, 101
fossil fuels, 16, 27–8, 42, 44–5, 99
Fossil fuels and the greenhouse effect, 44
Foster, Bob, 139, 140
Four Corners (television program), 3, 9, 13, 195
Fox TV, 198
France, 16, 25, 38
free-riding, 211
Frew, Wendy, 163
front organisations, 67, 129–30, 132, 138, 156, 158, 201–2
Frontiers of Freedom Institute, 67, 148
Furedi, Frank, 149

G (magazine), 111
G77, 31, 91
Garran, Robert, 74
Garrett, Peter, 152, 218
gas industry, 6, 16, 41, 95, 112
Gazard, David, 157
General Electric, 85n, 184
General Motors, 85, 130, 132
geosequestration, 125, 126
geothermal power, 18, 86
Germany, 25, 39, 40
Gilchrist, Gavin, 163
Gillott, John, 149
Global Climate Coalition, 83
global dimming, 181

Global Nuclear Energy Partnership, 205
global warming, 22–9, 109, 140, 181, 209–11
Goddard Institute for Space Studies, 181
gold industry, 41
goods and services tax, 99
Goodyear, Doug, 131
Gore, Al, 132, 150, 164, 180, 194–6, 198–200, 225
Grattan, Michelle, 60
Great Barrier Reef, 27
Greece, 79
green consumerism, 108–11
Green Party, 125
Greenhouse 21C package, 50
Greenhouse Challenge Program, 50–2, 97–8
'greenhouse effect', 27
Greenhouse Gas Abatement Program, 122
greenhouse gas emissions
  from aluminium smelters, 119–20
  in Australia, 36–43, *37–9*, 159
  climate change due to, 23–5, 29–30
  consideration of in policy, 49
  cost of reducing, 121–3
  European Union emissions trading system, 186
  industry opposition to reduction of, 49, 115–21
  inventories of, 36
  Kyoto Protocol targets for, 77–9
  from land-clearing, 80–1, *81*
  National Greenhouse Response Strategy, 48–50
  ONA report on, 44–6
  states adopt Toronto Target, 46
  US caps on, 183–6
  voluntary reductions of, 15, 50–4, 108–11, 160
  ways of reducing, 43
  see also carbon dioxide; fossil-fuel lobbyists; methane; nitrous oxide
greenhouse mafia, 3–12, 66, 116, 226
Greenland ice sheet, melting of, 180–1
Greenpeace, 68, 83, 160, 189

Greenpower, 52–4, 108
Griffin, Murray, 163
GST, 99
The Guardian, 149, 201
Gyngell, Allan, 209

Hagel, Chuck, 67
The Hague Conference (2000), 87–8, 121
Hannagan Bushnell, 67, 113
Hannagan, John, 74, 212
Hansen, James, 181
Hanson, Pauline, 140n
Hashimoto, Ryutaro, 58
Hawke, Bob, 45–6
Hawke Government, 46–50, 134, 158–9
Hawkins, David, 157
Haymet, Tony, 14
Hazelwood power station, Latrobe Valley, 120
heatwaves, 25
Hellicar, Meredith, 114–15
Henderson, Anne, 200
The Herald Sun, 198, 208
Herfurth, Welf, 140n
Hill, Robert
    attends Countdown to Kyoto conference, 68
    compared with Ian Campbell, 178
    denies possession of modelling, 72
    holds briefings for Australian journalists, 73
    insists on 'Australia clause' on land-clearing, 74, 77, 80
    leads delegation to Kyoto, 71
    loses trust of ministerial committee, 8
    not trusted to administer Kyoto deal, 97
    Pearse as former adviser to, 3
    power struggle over climate change policy, 98
    receives standing ovation, 96
    signs Kyoto Protocol, 99
    upbeat about Australian negotiations, 57
'hockey stick' debate, 21–2
Honda, 85

How to spend $50 billion to make the world a better place (Lomborg), 152
Howard Government
    adopts pro-nuclear position, 204–5
    announces commitments to industry, 102–3
    announces funding for solar electricity plant, 170
    anxious to keep seat at Kyoto table, 177
    appoints Batterham as Chief Scientist, 124–6
    appoints fossil-fuel lobbyists as Kyoto delegates, 6
    argues against signing Kyoto Protocol, 42, 100, 101
    attitude towards greenhouse gas emissions, 33, 38
    awards Morgan an OA, 135
    blames China as real emissions culprit, 175–6, 224
    claims Kyoto Protocol unfair, 31, 32, 100
    commissions new economic modelling, 103–4
    controls debate on greenhouse emissions, 13
    costings of compliance with Kyoto Protocol, 121–3
    denies seriousness of climate change, 13
    dismisses claims of sea-level rise, 92
    embarrassed over stance on Kyoto Protocol, 115
    eschews environmentalism, 159
    establishes AGO, 36
    exaggerates effect of adopting protocol, 68, 104
    feigns concern over climate change, 10–11
    finds excuses for inaction, 229
    hostility to wind farming, 165–8
    hosts CHOGM (2002), 91–2
    humiliated by observer status at meeting, 177
    ignores view of academic economists, 123

influence of fossil-fuel lobby on, 3, 6–12, 15, 18, 66
insists on right to trade emissions, 105
justifies lack of emission limits, 172
lies about climate change policy, 14–15
lobbies against emission reductions, 68
loses Australia's reputation as a global citizen, 55, 57–8, 76
ministerial greenhouse committee, 9
mounts case for differentiation, 55–8
negotiating strategy at Kyoto Conference, 70–5
overview of climate strategy, 221
power struggles over climate change policy, 98
promotes 'clean coal' technology, 225, 227
refuses to ratify Kyoto Protocol, 102
sabotages Kyoto Protocol, 221, 224
searches for loopholes in Kyoto Protocol, 87
sees Europe as the enemy re protocols, 64–5
shifts blame to developing countries, 100, 101
short-term energy strategy, 19
spreads fear among developing countries, 70–1
stifles protest in Pacific, 93–4
suffers climate change policy paralysis, 102
supports AP6, 187
in thrall to aluminium industry, 118
urged by Gore to change its position, 194
uses ABARE economic modelling, 60–4
voluntarism in emissions reduction policy, 108–10
Howard, John
announces climate change policy initiative, 68–9
announces Flannery as Australian of the Year, 220
announces task force on carbon trading, 212
appoints Morgan to RBA Board, 133
appoints Tuckey minister for forests, 178
approach to climate change, 190–1, 195
argues for Australia's 'special position', 42
briefed on Bush's nuclear plans, 205
consults polluters on greenhouse policy, 12
convenes meeting of LETAG, 10–11
convenes task force on nuclear energy, 206–7
describes AP6 pact as historic, 187
dictates greenhouse policy, 220
dismisses claims regarding bushfires, 216
entreated by academics to sign Protocol, 123
fossil-fuel lobbyist writes to, 50
ignores ONA report on emissions, 46
on *An Inconvenient Truth,* 195
pleads case for differentiation, 57
praised by Kelly, 214–15
refuses to accept advice of scientists, 216
supports nuclear power, 203–8
turns climate problem into water problem, 219–20
uninterested in greenhouse policy advice, 99
Howard, Lyall, 12
H.R. Nicholls Society, 133, 136, 139
Humane Society International, 166
Hunt, Barrie, 216, 217–18
Hunt, Greg, 206
Hunter Valley coal-washery power plant, 48–9
Hurricane Katrina, 179
Hurricane Wilma, 179

hybrid engines, 17, 85
Hyde, Tim, 131
Hydro Electric Corporation, 113
hydroelectricity, 16
hydrogen engines, 17
hydropower, 86

IBM, 84, 85n, 132
Iceland, 78
ICI, 51
Illarionov, Andrei, 156
income, 35
An Inconvenient Truth (film), 132, 150, 180, 194, 198, 199–200
Independent Television Commission (UK), 148
India
    declining crop yields, 92
    greenhouse gas emissions, 39
    joins AP6, 187
    Narmada Dam, 148
    participates in CDM, 215
    signs Asia-Pacific Partnership, 190
individual responsibility, 109
industrialised countries, 29–30
Institute of Public Affairs, 123, 139, 152, 155, 156, 157, 169
insurance costs, 34–5
insurance industry, 6, 86
Intercontinental Energy, 85
Intergovernmental Panel on Climate Change
    Second Assessment Report (1996), 22–3
    Third Assessment Report (2001), 23–4, 91, 92, 173
    Fourth Assessment Report (2007), 24, 127, 132, 196–7
    Bush dismisses as 'foreign science', 90
    formation of, 144
    overview, 20
    on protocol exemptions, 88
    research on climate change, 27
    science academies sign endorsement of, 182
International Aluminium Institute, 119

International Energy Agency, 43
International Policy Network, 202
International Tribunal for the Law of the Sea, 76
investment, in renewable energy, 17–18, 85–6, 87, 108
IPCC *see* Intergovernmental Panel on Climate Change
Irvine, Ross, 156–7
Italy, 38

Japan
    companies seek emission credits in Australia, 122
    greenhouse gas emissions, 39, 40
    joins AP6, 187
    Kyoto Protocol targets for, 78
    negotiations on tuna fishing, 76
    nuclear power in, 16
    opportunity for 'offshore compliance', 95, 106
    ratifies Kyoto Protocol, 94
    reduces reliance on fossil fuels, 42
    searches for loopholes in Kyoto Protocol, 87
    signs Asia-Pacific Partnership, 190
Jennings, Philip, 14
Johnson & Johnson, 84
Jones, Alan, 175, 216
Jones, Barry, 4, 102
Jones, Tony, 125, 213
Jorgensen, Peter, 76
justice, principles of, 31–4

Kakadu National Park, 134
Karoly, David, 145
Kavanagh, Trevor, 199
Keating Government, 6, 50, 159
Keating, Paul, 5
Kellow, Aynsley, 140
Kelly, Fran, 196
Kelly, Paul, 214–15
Kemp, David, 104
Kennett Government, 117, 121
Kennett, Jeff, 133
Kerr, Duncan, 75
Kininmonth, William, 20, 92n, 145–6, 200

Kiribati, 92, 93n
Knapp, Ron, 4, 120
Kohl, Helmut, 58
Kreger, Richard, 140n
Kyocera, 87n
Kyoto Conference (1997), 55–8, 70–7
Kyoto Protocol
    AGL support for, 112–13
    aims of, 30, 33
    attempts to undermine, 86, 221, 224
    Australia insists on seat at negotiations, 177–8
    Australia refuses to ratify, 102
    Australia's objections to, 29, 31, 42, 56
    Business Council support for, 114
    businessmen fear support of, 6
    China's obligations under, 175
    corporate support for, 85, 120
    corporations discuss how to scuttle, 137
    costings of compliance, 103–4, 121–3
    on developing countries, 100
    elements of, 77–9
    emission targets set by, 37, 78, 84
    Hill signs in New York, 99
    Latham promises to ratify, 10
    Lavoisier Group on consequences of ratifying, 141–2
    limited effect of, 27
    Marrakech negotiations, 103
    Morgan on repercussions of, 134
    Nairobi negotiations, 175–6, 210, 212
    the 'new' Kyoto, 188, 210, 215
    provisions for emissions trading, 105, 106
    public support for, 69, 210
    Russia decides to ratify, 94–5
The Kyoto Protocol ... (Illarionov), 156

Lahey, Katie, 115
Lamarck, Jean-Baptiste, 130
land-clearing, 36, 40, 71–2, 72, 77, 80–1, *81*, 87, 105
Landscape Guardians, 168

Lateline (television program), 125, 212–13
Latham, Mark, 10, 33
Latrobe Valley Generators, 113, 121
Laupepa, Paani, 93
Lavoisier, Antoine-Laurent, 139
Lavoisier Group, 92n, 135, 136, 138, 139–44, 145, 158
Lawrence Berkeley Laboratories, 43
Lawson, Dick, 137
Lawson, Nigel, 213, 214
lead contamination, 35
lead-free petrol, 154
Leipzig Declaration, 86
LETAG, 10, 14, 108, 167
Lewis, Steve, 189
Liberal Party, 133, 143, 169–70, 218
Lieberman, Joe, 85n
lifestyle changes, 227
lignite, 28
Limits of the Continental Shelf, 76
Lindzen, Richard, 91, 92n, 158
livestock, methane emissions from, 105
Living Marxism, 149
LM Magazine, 149
lobbyists *see* aluminium industry; coal industry; electricity industry; fossil-fuel lobbyists; oil industry
Lockheed-Martin, 85
Lomborg, Bjorn, 150–5, 201
Love, Geoff, 144
Low Emission Technology Demonstration Fund, 11, 170
Lower Emissions Technology Advisory Group, 10, 14, 108, 167
Lowy Institute, 209
Lunn, Stephen, 74
Luntz, Frank, 130

Macfarlane, Ian, 10, 11, 152, 194
Macken, Julie, 145, 163, 205
malaria, 25
Maldives, 155
Malthouse, Mick, 194
Mandatory Renewable Energy Target, 11, 98, 107–8, 166, 167
Maniates, Michael, 110

Manton, Mike, 144
Marrakech negotiations (2001), 103
Marshall, Andrew, 162
Marshall, George, 228, 229
Marshall Islands, 23
McCain, John, 85n, 190
McCain-Lieberman Bill (US), 183–4
McDonald, Meg, 64, 65, 96, 121
McDonald's Restaurants, 130
McGauran, Peter, 169
McKibbin, Warwick, 93n, 103, 123, 207
Meacher, Michael, 91
media, 73, 102, 161–4
MEGABARE economic model, 60–4
methane, 28, 36, 37, 105
Michaels, Pat, 67
Michaels, Patrick, 92n
Miller, Claire, 163
Minchin, Nick, 97–8, 155, 194
Minerals Council of Australia, 4, 8, 104, 121, 122–3, 136
mining industry, 16, 104
Mining Industry Council, 133
'missing-in-action' phenomenon, 6
Mitchell, Chris, 196, 199
Mobil, 62, 83
 see also ExxonMobil
modelling, economic, 60–4, 72, 121
modernism, 164
Monbiot, George, 129, 149
Montgomery, David, 137n
Montreal, 170, 185, 186, 188
Moody-Stuart, Mark, 83
Moonies, 85
Moran, Alan, 123, 169
Morgan, David, 114
Morgan, Hugh, 8, 114–16, 120, 132–6, 139, 145–6
Morning Bulletin, 104
Morrison, Jason, 216
The Moscow Times, 156
Munich Re, 187
Murdoch, James, 199, 201
Murdoch, Lachlan, 199
Murdoch press, 9, 22, 130, 150, 189, 196–203
 see also *The Australian*; *The Sun*; *The Times*

Murdoch, Rupert, 198–9
Murray-Darling Basin, 145, 219
Murty, Tad, 92n
Myer Foundation, 208

Nagle, Bill, 112
Nairobi, 175–6, 210, 212
Narmada Dam, 148
Napthine, Dennis, 170
NASA, 27
National Academy of Sciences (US), 90–1
National Greenhouse Advisory Committee, 47
National Greenhouse Response Strategy, 48–50
National Mining Association (US), 137
National Party, 133
National Press Club, 12
National Review, 155
National Tidal Facility, 92
National Trust of Australia, 168–9
natural gas industry, 112
Nature, 153
Naughten, Barry, 225
Nauru, 92
Nelson, Brendan, 203–4
Nelson, Roger, 138
neo-liberal ideology, 49
Netherlands, 23, 40
New Orleans, 179
New South Wales, 29, 46, 52–3, 72
New South Wales Farmers Federation, 208
New South Wales Land and Environment Court, 48–9, 172
New South Wales Minerals Council, 198
New York Times, 184
New Zealand, 39, 78, 93, 94
News Corporation, 198
Newspoll, 210
NGRS, 48–50
Nicholson, Peter, 189
Nine facts about climate change, 139
Nine lies about global warming (Evans), 139, 158

Index / 263

nitrous oxide, 28, 37
Norton, Gale, 90
Norway, 16, 40, 78, 79
Noske, Mike, 170
nuclear energy task force, 206–7
nuclear power, 16, 42, 45–6, 203–8

Oberthur, Sebastian, 76–7
Observer, 162
Office of National Assessments, 44–6
oil, extraction of, 28
oil industry, 16, 82–3, 85–6, 154
oil reserves, 30
oil shocks, 42
O'Keefe, Andrew, 194
O'Keefe, Bill, 137
Olson, Robert, 115
ONA, 44–6
O'Neill, Mark, 5, 9
O'Neill, Paul, 90
OPEC, 76
opinion polls, 69
orange-bellied parrot, 166–7
Orchison, Keith, 5
Organisation of Economic Cooperation and Development, 40, 43
Organisation of Petroleum Exporting Countries, 76
Orica, 10
Origin Energy, 10, 53
Osborn, Wayne, 12, 115
O'Sullivan, Paul, 137–8, 222
Ott, Hermann, 76–7
Oxley, Alan, 67, 140
ozone depletion, 110

Pacific countries, climate change in, 92–3
Pacific Island Forum, 94
Pacific Power, 113
Packer, Kerry, 33–4
Pakistan, 92
Paltridge, Garth, 20, 140
Parbo, Arvi, 132, 139
Parer, Warwick, 51, 59–60, 62, 102
Partnership for Climate Action, 84
Pattiaratchi, Chari, 174
Pearman, Graeme, 13, 139

Pearse, Guy, 3, 5–6, 7–8, 9–10
Pearson, Christopher, 213–14
peer review, 21
Pentagon report on climate change, 162
permafrost, melting of, 181
Perth, 174
Pew Center (US), 120
Pew Charitable Trust, 85
philanthropic organisations, 208
Philip Morris, 128–9
photolysis, 227
photovoltaic power, 18, 86, 227
Pittock, Barrie, 13, 20–1
Plimer, Ian, 20, 92n, 140, 189
Plomin, Robert, 149
Poland, 78
Polaroid, 84
Politics of fear ... (Furedi), 149
'polluter pays' principle, 32, 56, 58, 79, 100
Pomeranz, Margaret, 195–6, 200
Popper, Karl, 200
Portugal, 79
Powell, Alan, 63
Powell, Colin, 90
Prime Ministerial Working Group (on greenhouse gas emissions), 47
Prime Minister's Science, Engineering and Innovation Council, 124
public opinion, on climate change, 158–61, 209–10
Public Relations Institute of Australia, 157
Pulp and Paper Manufacturers Federation, 4, 97–8
Putin, Vladimir, 95

Queensland, 29, 72, 164
Queensland Coal Mine Management, 59
Quiggin, John, 61

Radio National, 196
rail transport, 17
Rann, Mike, 216
refugees *see* environmental refugees
religious groups, 208–9

renewable energy
    Australia's advantage in, 18–19
    Greenpower schemes, 52–3
    investment in, 17–18, 85–6, 87, 108, 227
    lack of sympathy for in Canberra, 66
    Mandatory Renewable Energy Target, 11, 98, 107–8
    in the National Greenhouse Response Strategy, 48
    responds to MRET, 11
    silencing of researchers of, 14
revenue recycling, 107
Revolutionary Communist Party, 149
Rice, Condoleezza, 190
rich man's indulgence syndrome, 111
Richards, Frank (pseudonym), 149
rights, 32–3
Rio Tinto
    attends anti-activism workshop, 157
    Batterham works as chief technologist at, 124, 126
    as donor to the Liberal Party, 133
    as a member of LETAG, 10, 12
    parent company support for Kyoto Protocol, 120
    supports Pew Charitable Trust, 85n
R.J. Reyolds Tobacco Company, 131
Robson, Alex, 123
Rothwell, Don, 177–8
Royal Dutch Shell, 82, 83
Royal Society, 202
Royal Society for the Protection of Birds (UK), 169
Ruddock, Phillip, 93
Russia
    as an exporter of energy, 41
    decides to ratify Kyoto Protocol, 94–5
    impedes progress at Kyoto, 76
    Kyoto Protocol targets for, 78
    melting of permafrost in, 181
    searches for loopholes in Kyoto Protocol, 87

Saeed al-Sahaf, Mohammed, 171
Safeguarding the Future: Australia's response to climate change, 68–9
Samuel Griffith Society, 136
Sandalow, David, 190
Santoro, Santo, 155
Saunders, Peter, 150
schistosomiasis, 25
Schwarzenegger, Arnold, 185
science
    agreement among climate scientists, 22–7, 195–6
    corporate acceptance of climate science, 82–3
    denials of climate science, 20–2, 92n, 123
    Lavoisier Group on conspiracy by climate scientists, 141
    politicisation of, 13–14
    process of peer review, 21
    role of climate science, 19–22
    science academies sign endorsement of IPCC, 182
Scientific American, 153
Scripps Institution of Oceanography (US), 14
sea-level, rises in, 173–4
Senate Foreign Relations Committee (US), 91
Senate Inquiry into Global Warming, 71n
Senate Estimates Committee, 51
7.30 Report (television program), 219
Sharp, 87n
Shearer, Ivan, 76
Shell Oil, 51, 83, 85, 86, 157
Schipper, Lee, 43
Siemens, 87n
Silent Spring (Carson), 151
Simon, Julian, 151
Singer, Fred, 92n, 148, 158
Sinodinos, Arthur, 9
60 Minutes (television program), 152
The skeptical environmentalist (Lomborg), 150–2
Smith, R.J., 137, 138
smoking, 128–9
snail fever, 25

Snowe, Olympia, 185
Snowy Mountain Scheme, 113
Society for Egomaniacs, 6
solar power, 17, 19, 87n, 227
Soon, Willie, 158
South Korea, 187, 190
South Pacific Forum, 93–4
Spain, 25
The Spectator, 155
Spiked Online, 150
Staley, Tony, 140
Stanwell, 113
State of fear (Crichton), 157
stationary energy, 38, *38*
steel industry, 41
Steffen Report, 173–4
Steffen, Will, 173, 193
Stern, Sir Nicholas, 211
Stern Review (UK), 207, 211, 213, 216
Steyn, Mark, 155
Stockdale, Alan, 117
Stone, John, 68, 133
sugar industry, 108
The Sun, 199
Sun Oil, 82, 85
Suncor Energy, 84
The Sunday Age, 193
Sun-Herald, 164
surveys, 161
sustainable energy industries, 114
Sustainable Energy Industry Association, 114
Switkowski, Ziggy, 206, 208
Switzer, Tom, 199
Sydney Institute, 200
Sydney Morning Herald, 11, 57–8, 69, 151, 158, 163, 206–8
Synhorst, Tom, 131

Talake, Koloa, 92
Tambling Review, 11
Tasman Institute, 5n, 138
Tasmanian Conservation Trust, 166
TASSC, 128–9
Tate, Alan, 163
tax minimisation, 33–4
Tech Central Station, 130, 202
Tech Central Station Asia-Pacific, 67

Texaco, 62, 82, 83, 84
Thatcher, Margaret, 213
thermohaline circulation, 179, 181
Thomson, Andrew, 143
'three mines' policy, 206
3M, 85
tidal power, 18
Tilley, John, 5
Time, 151
The Times, 200
'tipping points', 179–82
tobacco industry, 128–9, 131
Tonga, 92
Toronto Target, 46–7
tourism industry, 6, 164
Toyne, Phillip, 6
Toyota, 17, 84, 85
trade, 41, 228
transport, greenhouse gas emissions from, 38, *38*
Tuckey, Wilson, 178
tuna fishing, 76
Turnbull, Malcolm, 175, 178, 218–20, 225
Tuvalu, 92, 93
2GB, 216

Ukraine, 78, 95
Umbrella Group, 64, 87
UNFCCC, 32, 47, 56, 100, 177, 215n
Union of Concerned Scientists, 202
United Kingdom, 25, 35, 39, *39*, 58
United Kingdom Meteorological Office, 27
United Nations Conference on Environment and Development, Rio de Janeiro, 47
United Nations Framework Convention on Climate Change, 32, 47, 56, 100, 177, 215n
United States
  air quality, 35
  criticises Kyoto Protocol, 31, 32, 89
  denial of climate change, 128–32
  greenhouse gas emissions, 39, *39*, 40
  joins AP6, 187
  Kyoto Protocol targets for, 78

McCain-Lieberman Bill (US) on emissions, 183–4
oil companies, 83–4
opportunity for 'offshore compliance', 95, 106
Pentagon report on climate change, 162
proposals to cap greenhouse emissions, 183–6
refuses to commit to emissions reductions, 225
refuses to ratify Kyoto Protocol, 95
searches for loopholes in Kyoto Protocol, 87, 88
Subcommittee refuses funding of Asia-Pacific Partnership, 192
urban pollution, 154
Utah Mining, 59

vehicle efficiency, 40, 84–5
vehicle emissions, 84–5
Victoria, 28, 46, 169
Victorian Department of Treasury and Finance, 117
Vietnam War, Ray Evans on, 139
Virgin Galactic, 111
voluntarism, in emissions reduction, 15, 50–4, 108–11, 160

The Wall Street Journal, 201
Wallop, Malcolm, 67
Walsh, Peter, 5, 140, 142
Walsh, Sam, 10, 12
warming *see* global warming
Warren, Matthew, 189, 198, 200, 223
The Washington Post, 82, 85, 151

water policy, 219–20
water restrictions, 193
Watson, Harlan, 188
Watson, Robert, 22–3
The weather makers (Flannery), 109
Webber, Ian, 139
Wells, Dick, 4, 8–9
Wesfarmers, 133
West Antarctic ice sheet, 180
The West Australian, 174
Western Australia, 46, 174
Western Mining Corporation, 67n, 116, 132–3, 136
Westpac, 114
Whirlpool, 85
Wikipedia, 132
Wilson, Kath, 157
wind power, 11, 18, 53, 86–7, 108, 165–9, 227
Wind Power Pty Ltd, 166, 167
Windschuttle, Keith, 136
Wirth, Timothy, 57, 137
Wise Use movement, 148
Wood, Alan, 152, 158, 202–3, 214
Woodford, James, 57–8
Woodside, 8, 115
World Coal Institute, 4
World Meteorological Organization, 144
World Summit on Sustainable Development, 94
World Trade Organization, 95
WWF-Australia, 160, 166

Zillman, John, 144–5